The Darwin Conspiracy

Acknowledgments

MY DEBT to Paul Hannon is not easily quantifiable – for his enthusiasm, generosity and his determination in the pursuit of evidence; also to Dilys Allam, then at the publishers Taylor & Francis, for her belief, commitment and energy at a crucial time; George Beccaloni for honest appraisal, constant interest and a shared belief in the importance and genius of Alfred Russel Wallace; Jim Moore for excellent advice offered despite obvious misgivings; Andrew Berry for recognising a gap in Darwin studies and for offering the encouragement and personal introduction which gave me the confidence to complete the work; Janet Browne for making time in a hectic relocation between London and Harvard to talk to me about ideas, Wallace and Edward Forbes; Harry Dean for his open-mindedness in changing from initial sceptic to enthusiast and for his continuing encouragement and advice throughout; Femme Gaastra whose knowledge and expertise has illuminated with absolute certainty the shipping schedules between the Malay Archipelago and England in the mid-nineteenth century; Matthew Davies and Eugene Weber whose comments and suggestions helped in so many ways; Elin Rhys of the Welsh television company Teledu Telesgop who with a mixture of encouragement, challenge and humour sustained the development of this book in the dark days when many were absorbed by the pitch but so few felt the need to make it visual; and finally my editor Rowan Davies who turned a difficult concept into an infinitely more accessible manuscript.

I wish also to recognise the help given me by the late Colin Paterson of the Natural History Museum in London who was prepared to listen when I needed it most, who read the earliest outline without hesitation though we had never previously met, and who liberated Dov Ospovat's book from the Museum's bookshop as he passed it a few moments before our first meeting, and to John Langdon Brooks, who also died during the research stage of this book, who criticised and helped after reading my initial attempts at a summary of this story and who gently warned of the obstacles I was likely to face further down the road.

Others have been continually generous with their time, help and advice especially Gina Douglas and her colleague Lynda Brooks at the library of the Linnean Society and members of staff of the Zoological Library of the Natural History Museum. I am also indebted to the Cambridge University Press for permission to publish extracts from *The Correspondence of Charles Darwin* and to the Syndics of Cambridge University Library for their own permission to use extracts from their archive and for their timely decision to make Darwin's correspondence accessible online. The Linnean Society generously gave permission to publish extracts from the journals and notebooks of Alfred Russel Wallace as did the American Philosophical Society regarding Darwin's letters to Sir Charles Lyell. Finally I would like to thank the many other publishers who allowed extracts from their authors' books to be used as material evidence in the structuring of this story.

Introduction

IN THE mid-nineteenth century, Great Britain boasted two scientists who were to change forever the way mankind viewed the world. Despite their individual gifts, they were as unlike as any two men of ideas could possibly be. One was rich, privileged, highly educated and connected to important families in Victorian society; the other worked for a living, had no significant social connections and left school at the age of fourteen.

The first, after years of hunting and shooting, became conventional and dutiful, and bore the weight of expectation willingly; the other, self-taught, was instinctive, radical, free-thinking, open and unbound by convention throughout his life. The older figure, following a long voyage of wonder in his young manhood, married, bred a large family, inherited substantial wealth and invested wisely, but afterwards rarely travelled far from his home in rural Kent in the southeast of England. The younger, who had no thoughts of marriage, roamed, often alone, the rainforests of the Amazon basin and modern Malaysia and Indonesia. One was brought up to believe in and respect the Christian Church (and even, at one time, to consider it as a career), while the other had no time for its dogma and paralysing power over the minds of men.

Of one, Charles Darwin, you will most certainly have heard. Of the other, Alfred Russel Wallace, you may well have heard nothing.

One hundred and fifty years ago, after exhausting feats of research, observation and analysis, both men were acknowledged by their peers to be considered joint discoverers of the theory of evolution. At that time, the stranglehold of religion on Victorian society was such that only brave or foolish people rejected its teaching and its authority. Along with everyone else, both men were expected to accept without question that everything in the world was the work of a knowing, ever-present, all-powerful God. The Church told people how to act, what to think, when to work and when to rest.

The influence of the Church had not always been this great. In former times, philosophers, artists and wise men, from Ancient Greece to Renaissance Europe, had pondered how the incredible variety of forms of life they saw around them might have been produced. Some came within touching distance of what we now know as the theory of evolution. However, although many argued for the fact of evolution, none could indicate exactly how it happened.

The theory of natural selection has been defined as 'a bias acting within a species which promotes the survival of some variants and not others thus making for a change within the species'.[1] The statement is lucid and simple, but the solution to the essential question – how does that initial bias become so magnified that it can cause change? – eluded many of the greatest thinkers for more than two thousand years. Ideas of causal factors swung between the work of a Creator and nature's inherent patterning. Empedocles, Aristotle, Epicurus, Lucretius, Gregory of Nyssa, St. Augustine, Avicenna and Leonardo da Vinci had all offered their thoughts before Francisco Suárez (1548–1617), a Spanish Jesuit priest writing after the Moors had been banished from Spain, attempted to put an end to secular speculation and declared that every form of life was created by God, and remained unchanged until it became extinct. He also insisted that no living organism could mutate into another form. Suarez's doctrine became the official policy for the Christian Church, and was a direct cause of the strict orthodoxy that was still in place at the beginning of the nineteenth century.

In the intervening centuries, Francis Bacon (1561–1626), René Descartes (1596–1650) and Gottfried Wilhelm Leibniz (1646–1716) all favoured natural causes for species change, before Carl Linnaeus (1707–1778) insisted that species are created by God, and are forever fixed and unchanging. His direct contemporary, Georges-Louis Leclerc Buffon (1707–1788), while he was unable to dismiss God's influence entirely, invoked natural causes; even Charles Darwin's own grandfather Erasmus (1731–1802) claimed, on little evidence, that the ape was a direct predecessor of man. In the two volumes of his *Zoonomia* published in 1794 and 1796, he also expressed the idea that characteristics acquired by individuals in the struggle for life could be passed on to their descendants. A few years later, Jean-Baptiste Lamarck (1744–1829) argued that animals mutate because the struggle to survive causes advantageous changes in physical structure, which can be passed on to offspring. This idea left Lamarck open to criticism and eventually undermined his arguments. The way was then left clear for Étienne Geoffroy St. Hilaire (1722–1844) to suggest that some organisms adapt and survive when environments change, while unsuccessful adaptations perish and become extinct. St. Hilaire had come closest to understanding the process of evolutionary change, but he, also, failed to explain exactly what caused an animal to change its form, or why millions of species on earth had emerged from a common ancestral species. After St. Hilaire's pronouncement, Georges Cuvier, a French academic, criticised both Lamarck and St. Hilaire for theorising, rather than dealing with the facts of the natural world around them. In the meantime, Alphonse de Candolle, Jean Louis Agassiz and Charles Lyell offered alternative ideas, which again led the emphasis of the debate back to the omnipotence of a Creator.

By 1830, the stage was set to welcome the incredibly fertile brains of Charles Darwin and Alfred Russel Wallace, who, ten years apart and unknown to each other, were to undertake the challenge of solving the mystery of the origin of species. They both used Charles Lyell's ideas as a springboard. Because of their backgrounds, however, they were to take entirely different elements of his convictions as starting points.

Charles Lyell (1797–1875) was a deeply religious man who was convinced that everything in the world had been created by God. He was also a brilliant natural philosopher who helped to provide convincing geological evidence that the planet was millions of years old, and had certainly not been created little more than four thousand years before Christ, which was the orthodox teaching of the Church. He managed to reconcile the differences between his religious belief and his geological discoveries to his own satisfaction, but both Darwin and Wallace were to reject the elements of Lyell's theory that did not appeal to their background beliefs, and accept those elements that fitted their needs.

Darwin's early theoretical approach was based partly on Lyell's religious belief in a world of harmony and balance created by God, where species were perfectly adapted to their environment. Beyond that point, however, he had no need of a supernatural force in his explanation of natural selection. Wallace, a few years later, shunned Lyell's belief in a world devised by a Creator, but grasped with both hands Lyell's geological discoveries and used them as primary evidence that evolution had no need of God. He discounted all Lyell's religious ideas, but was fascinated by his theories of geological strata and evidence of fossils found in sequences of rock layers. He drew strength and certainty from Lyell's practical teachings about the real age of the world and folded these ideas into his own development of a theory of evolution.

Years later, both Charles Darwin and Alfred Russel Wallace were acclaimed equally for unlocking the secrets of the evolutionary process. So how did Darwin manufacture so much fame for himself and how has Wallace been denied his place in the pantheon of great British scientists so completely, despite the fact that they were both credited at the same time and on the same day with having discovered one of the most important truths about the natural world? The answers are to be found in the communications system of the mid-nineteenth century, something that is of great significance in this tale of Darwin's rivalry, ambition and subsequent plagiarism.

In the main period covered by this book, communication was by letter. The written word was threatened only by the new technology of the telegraph, which linked countries but not yet continents, and transmitted only the briefest messages. Letters delivered the news, bonded relations and kept empires functioning. Backed by the British Government and the Admiralty, the system of despatching and delivering carefully guarded shipments of letters, packages and parcels around the world was, accidents notwithstanding, totally secure and incredibly efficient. 'The mail' was the news. It came to Britain by ship from the United States, Canada, the West Indies, South America and South Africa. Twice every month, the India and China mails came from the Far East in secured boxes. These shipments were under the strict supervision of former Royal Naval officers, who had the authority to overrule the ships' captains if the safety of the mail was threatened.

For the political leaders of a country engaged in extensive military actions overseas (there was unrest in India and attacks on British naval forces in China), the latest reports from ambassadors, Army and Navy commanders and government officials were crucially important. For financial and general news, there was the background reporting and interpretation of events by correspondents of *The Times*, whose reports were completed and taken aboard even as the homeward-bound liners readied themselves to leave port. Short telegraph messages from ports in the Mediterranean gave hints of longer reports contained in the first-class mail that would arrive in London by boat-train within a day. Then, about five days later, eyewitnesses to unfolding events in Bengal or Canton landed in Southampton after a six-week journey on the high seas. For those in England who waited in expectation, the arrival of the bulk mail and the news it delivered made the headlines.

The weather sometimes slowed the progress of a ship, but usually only by a day or two. The timetable was strictly observed, and letters went astray so rarely that it is difficult to find any examples. The Dutch, who used the same Peninsular and Oriental (P&O) shipping liners as the British Government, reported that in the whole of 1858, of the myriad letters received via the India and China mails, only one was delivered late at the house to which it was addressed.

The delivery of the India and China mails is an immensely important detail in the apparent coincidence of Darwin and Wallace cracking the evolutionary code at precisely the same time from such different starting points. The story that will be told in this book is light-years away from the established orthodoxy, which states that a letter from Wallace caused Darwin to rush to establish his claim to be the first to outline the theory of evolution. An increasing body of evidence contradicts the received view of Charles Darwin as a benevolent man who, alone, unaided and without precursors, was inspired to write *On the Origin of Species*. At the heart of that famous historical event lies a deliberate and iniquitous case of intellectual theft, deceit and lies perpetrated by Charles Darwin. This book will also argue that two of the greatest Victorian scientists were willing accomplices.

Some academics who have studied the development of Darwin's theory have found glaring inconsistencies that are not easily answered. This research has dripped steadily but infrequently into the literature surrounding the achievements of Darwin and Wallace over the past fifty years. Assessed collectively and dispassionately, it presents an astonishing glimpse into the mind and motivation of a man claimed as Britain's greatest natural scientist. As well as presenting contemporary events in the lives of Darwin and Wallace and their fellow scientists, this book traces the progress of this historiographic detective work.

In his lifetime, Charles Darwin wrote some fourteen thousand letters, and came to know the shipping timetables very well. Now published as *The Correspondence of Charles Darwin*, his letters form a record that is unlikely to have any equivalent in today's electronic world. The efficiency of the Post Office, and Darwin's meticulous filing system, means that he can be continually assessed and reassessed, nearly 130 years after his death.

It was, therefore, inevitable that Darwin's life would be examined closely by academic researchers. Among papers that were first made available fifty years ago, scholars, for the most part American, began to uncover anomalies and coincidences that allowed them to question Darwin's probity and ethical behaviour in the years following his return from the *Beagle* voyage. Rapid advances in photocopying and microfilm technology during the 1950s allowed some of them access to documents they needed in order to verify their claims.

From the facts they unearthed, supplemented by new evidence discovered while researching this book, there is little doubt that a compelling case can be made against Darwin that would allow any reasonable person to conclude it is likely he committed one of the greatest thefts of intellectual property in the history of science.

CHAPTER 1

Windfall

IN THE late summer of 1856, Charles Darwin was under tremendous pressure to solve what was commonly termed 'the species question': namely, how and in what circumstances did new species come into existence. After nearly eighteen years of thought, he found himself stranded. He had laboured over a theory he had hoped would explain how one species might change into another, but he was tiring in the face of cogent and persuasive evidence from friends and colleagues, who insisted that a core element of his theory was wrong. To his closest confidants, he sometimes admitted that he was completely baffled by the appearance of varieties and the origination of new species in the absence of a change in environmental conditions.[1]

Frustrated by his inability to solve the problems at the heart of evolutionary theory, Darwin had very few places left to go. Four months earlier, he had given Sir Charles Lyell, his friend and mentor, an assurance[2] that he would think about writing an outline of his ideas for publication. Left to himself, it was not a course Darwin would have chosen,[3] but in late 1855[4] Lyell had read an article that indicated that Darwin had a new rival in the race to be recognised as the first man to understand and explain the secret of evolution. The article (entitled 'On the law which has regulated the introduction of new species' and published in the *Annals and Magazine of Natural History* in September 1855) had so impressed Lyell that he had opened a notebook himself, and had written on its first page not the name of Charles Darwin, his protégé, but that of Alfred Russel Wallace,[5] a man who made his living by catching beetles, birds and butterflies in the island jungles of the Malay Archipelago, ten thousand miles away.

Darwin had also read the article, but had dismissed it (in notes written in the privacy of his own study) as communicating nothing of great interest.[6] However, when he visited Darwin at the latter's home in Kent (in the countryside outside London) in April 1856, Lyell insisted that Wallace's ideas were so revolutionary that they were a direct threat to

Darwin's long-coveted ambition. Wallace had begun to believe that new species were to be found only in environments in which closely-related species had previously existed. Lyell found this idea very convincing but it contrasted sharply with Darwin's belief that new species could originate only in newly-formed and isolated environments like islands to which already existing organic forms had migrated. Within three weeks of Lyell's warning, in an attempt to protect what he regarded as his scientific priority, Charles Darwin began to outline his ideas about how species evolve from one form into another.[7] He gave the manuscript the title 'Natural Selection'. Darwin had previously made a study of barnacles, and this had given him a clue that all species to some extent give rise to new varieties.[8] However, the idea of variation had not become central to his thinking. His long-held conviction (first expressed by Lyell) remained unchanged: he believed that the already-populated world was a place of balance and harmony, every organism was perfectly adapted to its environment, and species existing there could not mutate from one form into another.[9]

Despite the fact that several of his friends, including Lyell, had expressed serious doubts about his ideas concerning species migration, Darwin was convinced he was right, and that sooner or later the sheer weight of the facts that he was collecting from all over the world would prove it. Yet he could not explain why species existing in conditions of perfect balance and harmony should produce varieties with distinct differences from the parent form.[10] Perhaps he was hoping that the final link in the chain would become clear, possibly from his interest in pigeon breeding. As 1856 drew to a close, he was still far from completing his manuscript, and thereby becoming the first person to reveal how new species originate.

Then, on 11 January 1857, the Post Office agent aboard a train bound for London took charge of bundles of second-class mail from the Far East. These letters and parcels had been carried to the port of Southampton by the Peninsular & Oriental steamer *Colombo*, which had docked earlier that day.[11] Among these bundles was a letter addressed to Darwin from Wallace, who was at that time based on the island of Celebes in the Malay Archipelago, between the immense land masses of Borneo and New Guinea.

In October 1856, when he wrote his letter to Darwin,[12] Wallace had begun to see that some connections between animals could be accounted for by a process of divergence and modification. This process of divergence could, in turn, be linked firmly to a continuing process of extinction.[13] Neither of these ideas had yet appeared in any of Darwin's writings. However, Wallace did not yet have any idea about what caused organisms to change their form and diverge from their ancestral species.

The purpose of his letter to Darwin was not to display his originality. Wallace certainly had no prior knowledge of Darwin's longstanding attempt to solve the species question. It was simply that after more than two years in the jungle, Wallace was yearning for intellectual stimulation and an exchange of ideas. Darwin must have seemed a man likely to be interested in some of Wallace's most recent discoveries.

The text of that letter from Wallace has never been revealed. Whatever the innocent Wallace had intended to convey to Darwin, when the letter arrived at Down House on 12 January 1857 it delivered his most radical insights straight into the hands of the man who needed them most. The contents of the letter have never been seen by anyone other than Darwin, but we do have Darwin's reply.[14] On 1 May 1857, nearly four months later, he wrote to Wallace claiming that he had received his letter only 'a few days ago' and he could see that he and Wallace had thought much alike. He explained to Wallace that he was writing a book about his findings and after nineteen years' work he hoped to publish something in the near future.

Charles Lyell's warning to Darwin one year previously, in April 1856, had been timely, but Lyell could have had no idea how quickly Wallace would develop his ideas, or how interesting they would become in the space of a year. He had known that Wallace's ideas were a threat to Darwin's ambition. In contrast to Darwin, the young butterfly collector was not intellectually constrained by the idea of a world designed by God so that all species existed in perfect harmony and balance, and where every species was perfectly adapted to its environment; nor did he feel that he had to make room for God in his theory. He had also had the chance to observe and compare animals and plants in the wild tropical rainforests of South America and the Malay Archipelago, ten thousand miles apart, and his ideas on species and varieties were influenced by what he had observed in both. Lyell recognised the certainty with which Wallace wrote about his ideas. However, Lyell had once led the field of geological theory himself, and he knew how difficult it was to get controversial ideas accepted.

In the early nineteenth century, when Lyell was actively researching, geology was regarded as a dangerous science. As such, it both attracted and repelled the public, who still believed that God had made the Earth, and everything on it, in less than one week. Theological authority was strong and there was the greatest pressure upon geologists to avoid direct conflict with the Church. Moreover, many of the new professional geologists were theologians (as was Lyell himself), and they were at great pains to reconcile geology with their religious beliefs. They were also confronted with the theory of catastrophism, whose advocates claimed that mountain ranges had not taken millions of years to form, but had been thrust up overnight. When Lyell was working out his own theory, he could have had no idea that one day he would be challenged by a man who had no need of God.

CHAPTER 2

The man who stretched time

WHEN the 31-year-old Charles Lyell explored the slopes of Sicily's Mount Etna in the summer of 1828, he had been looking for evidence that incredible geological forces operating since the beginning of time had constantly moulded and remoulded the surface of the planet. He soon discovered that in one place ancient seabeds had been raised more than 700 feet above sea level.[1] This fact among others convinced him that Mount Etna was vastly older than anyone could imagine. A few weeks later in the south of Sicily he noted a bed of fossilised oyster shells lying on top of a bed of lava twenty feet thick, which itself had been completely covered by another lava bed.[2] He realised the fossilised oysters had probably been overwhelmed by a second lava flow from Mount Etna while still alive and feeding on the sea floor. This must have occurred before the entire ancient seabed was pushed upwards to become dry land.

Lyell believed that in any environment some species would become extinct and other species would take their place. All the 'original' species would be replaced by entirely new species over a long period of time. One thing bothered Lyell: in an environment in which geological forces had made all species extinct, how did an entire population of new species come into being? His biographer put it simply:

> when an island like Sicily first rose from the sea, as Lyell now believed it had risen from the sea, it would be a bare surface with no native species or plants of its own. Consequently, any plants it acquired would be those which could colonize it from neighbouring areas of land. Hence arose the paradox – which so much struck Lyell when he contemplated the geological history of Sicily – that the plants growing on the island belonged to species much older than the island itself.[3]

Had the plants growing on Sicily migrated there across the seas from other locations like Italy and Africa, where they had come into being at some 'centre of creation', or were plants created simultaneously at several different locations, all at the same time? Both ideas had a lot to be said for them.

Lyell had thought it possible that species had been created in succession 'for an appointed time', but that all the principal families of animals had existed since the dawn of time. Species became extinct only through the absence of food, the action of natural geological forces or climatic change, or because of new predators. He also believed that new species were deliberately designed to be perfectly adapted to certain environments, in order that they might populate newly created regions. New species were created by a 'Divine Hand' (he was never more specific than this). Each new species was modelled on one that already existed, ensuring a harmony of design in the biological community. In environments in which every species had become extinct, whole communities of species were created in replacement. These species fitted precisely into the gaps left by extinct species. This was to become known as Lyell's 'theory of multiple creations', a process by which the balance of nature was continually maintained.

In February 1829, Lyell broke his journey back to England to visit Augustin Pyrame de Candolle, a Swiss professor of botany, to talk over his ideas. De Candolle had long studied plant species and their geographical distribution, and also believed in the struggle for existence between species. He was convinced that the distribution patterns he had identified in plants could be accounted for only by species emerging from one single centre of creation.[4] Lyell later wrote:

> I am now convinced that geology is destined to throw upon this curious branch of enquiry and to receive from it in return much light, and by their material aid we shall soon solve the grand problem [of] whether the various living organic species came into being gradually and singly in insulated spots or centres of creation, or in various places at once and all at the same time. The latter cannot, I am already persuaded, be maintained.[5]

Lyell had become enthusiastic about de Candolle's belief in the struggle for existence. De Candolle portrayed animals as existing in a state of war; those who died were the weakest fighters and those who survived were the strongest. This struggle mattered because Lyell's geological discoveries – and the fossil record – had convinced him that many species that once existed had become extinct. Lyell had been influenced by the arguments of Thomas Malthus, who wrote that population increase was limited by the amount of food available to feed a population. Lyell had been convinced by de Candolle that when it became difficult to procure sufficient food for the young of the species, or when a species became vulnerable to a predator, that species would fail to thrive and would eventually become extinct.

On his return to London after meeting de Candolle, Lyell began writing the first volume of his *Principles of Geology*. His assured style and the weight of evidence he had gleaned from his European travels ensured that it made a huge impact on the study of natural philosophy when it was published in 1830. It was the most thorough overview of geology ever written. It became central to the debate surrounding a question that had remained unresolved for more than thirty years: how had the earth's surface been formed? Lyell had become convinced by the theories of the Scottish geologist James Hutton.[6] Hutton had proposed that the surface of the earth was being continually re-shaped as land was pushed up from beneath the seabed and then eroded. Realising that such processes took place over considerable periods of time, Hutton bravely went against the consensus. In fact, he had argued (wrongly) that the earth was infinitely old. (His followers later settled on an age of several million years: still a huge underestimate by modern standards, but a significant step forwards in understanding.)

By the end of the nineteenth century, Hutton's theories had been widely accepted – indeed, in some ways they were accepted too completely, leading many to dismiss the modern theory of plate tectonics and to underestimate the role played by catastrophic events (such as asteroid strikes) in mass extinctions. However, when he proposed these ideas in the last decade of the eighteenth century, Hutton had absolutely no way of proving them and received little initial support. His research remained largely ignored until Lyell's interest was awakened.

Lyell's Italian expedition, which had thrown up evidence that the world was far older than had previously been believed, forced him to confront those who still argued that the world was only six thousand years old. 'Catastrophists', as they were known, were convinced the earth had experienced several severe and sudden catastrophic events in its extremely short history: the biblical Flood, gigantic tidal waves, earthquakes and paroxysms of the earth's crust. These cataclysmic events caused the extinction of every living creature. Following these extinctions, God would create new populations of species and plants, replicating those that had previously existed and setting them in the places they had previously occupied. The only difference was that the species created each time would be a little more advanced. Each successive creative act, therefore, drove steadily towards the perfection of man.

Since Hutton had proposed his theories at the end of the eighteenth century, the new science of palaeontology had fostered the discovery of differences in life-forms between successive geological eras. This seemed to indicate that there was no necessary progression between fossilised animal forms found in one rock stratum and those found in the next. Spurred on by this, Lyell rejected the Catastrophist theory that God had acted to create new but connected populations of organisms at various stages. 'We are not authorised in the infancy of our science to recur to extraordinary agents', he argued.[7] He knew there had to be an explanation for both geological and organic change on the changing surface of a world millions of years old, but as he came to terms with the complexity of the subject he desperately needed a scientific rather than a religious explanation.

As a result, Lyell came close to spelling out his own version of the principle of natural selection (although he did not use the term). Even though his definition was not stated in the clearest terms, he made the first systematic attempt to consider the factors affecting the extinction of species and the effects of climatic change upon animal life throughout the long course of ages:

> Every species which has spread itself from a small point over a wide area must have marked its progress by the diminution or the entire extirpation of some other, and must maintain its ground by a successful struggle against the encroachments of other plants and animals.[8]

Lyell called this his 'principle of preoccupancy'.[9] However, he viewed the struggle involved as a force of nature working to prevent species change rather than encouraging it to happen. He did not believe the theories concerning the transmutation of one species into another proposed by natural philosophers like Jean-Baptiste Lamarck. Citing his preoccupancy principle, he wrote:

> It is idle to dispute about the abstract possibility of the conversion of one species into another, when there are known causes so much more active in their nature which must always intervene and prevent the actual accomplishment of such conversions.[10]

Lyell knew that new species appeared on earth, but he could not imagine how they could come into being other than by the hand of a Creator. He believed that the already-created, the already-fit, dominated every corner of the habitable world. As geological conditions altered over time, new environments were formed and existing species moved to these virgin areas in order to colonise them. Lyell could see no evidence of a mechanism other than the hand of God to explain the emergence of new forms in those areas of a world already dominated by existing species. He refused to believe in the possibility of the transmutation of species.

Over the years, Lyell's geological arguments won the day and the theories of the Catastrophists were gradually rejected. However, it was not accidental that when Charles Darwin returned from the *Beagle* voyage in 1836, Lyell was quick to offer himself as a mentor and friend to the young man. Darwin's *Beagle* experiences, particularly his first-hand observation of a piece of Chilean coastline that had been suddenly elevated by a recent earthquake, enabled him to support Lyell's geological ideas. At the same time, Lyell's influence in the world of geology gave Darwin access to the inner circles of the new science and the beginnings of a bright future as a natural philosopher.

The two men needed each other. In Lyell, Darwin saw a protector, a guide, a powerful friend. In Darwin, Lyell saw more than a young acolyte. He invested heavily in Darwin's reputation. Darwin, in turn, adopted Charles Lyell's ideas as solid foundations while immersing himself in the idea of transmutation. Darwin believed that the adaptation of organisms to their environments was perfect, that nature was a well-adjusted mechanism, that there was harmony between organisms and that the laws of nature had been established by God to achieve His ends.

It was to be twenty years before Darwin rejected the assumptions that were deeply embedded in these traditional views.[11]

CHAPTER 3

Formation of a naturalist

BY THE end of 1845, Charles Darwin was famous around the world. Earlier that year he had published, for the second time, the journals that recorded his experiences aboard the British survey brig HMS *Beagle*. The voyage had ended nearly ten years before, but the earlier publication of his diaries written in 1839[1] had not been particularly successful: they had been published alongside other accounts of the naval expedition to survey the Atlantic and Pacific coastlines of South America.

The 1845 version was an entirely different commercial proposition.[2] For an audience eager for stories about unexplored lands, the book had everything. In more than a thousand pages, Darwin recounted nearly five years of adventure, danger, surprise, fascinating discoveries, strange animals and science. He told stories of the human tragedies of earthquakes, riding with gauchos, the treatment of slaves, and the strange customs and stranger ways of life he encountered as the *Beagle* made its halting progress from the tropics to the edge of the Antarctic and back home again. Darwin's easy writing style allowed readers to view him as someone brave, resourceful and inquiring, and reading the book was commonly held to be an enthralling experience.

In addition to all this, Darwin claimed that the journey had given him insights into one of the greatest scientific questions of the day: how and when do new animal species come into existence? These insights, based on factual evidence and claims of personal observations (particularly with regard to the animals of the Galapagos Islands) made the book more than an adventure story. It presented Charles Darwin as a scientist already actively engaged in the search for the answer to the origin of species.

The book was a phenomenon. It was both a professional triumph and a personal vindication for Darwin. It gave him fame and recognition, and it also conferred an aura of scientific respectability. This was a long way from the dissolute fate his despairing father had once predicted for him.

Born on 12 February 1809, Charles Robert Darwin was the fifth child of Robert and Susannah Darwin. His birthplace was their home, The

Mount, which overlooked the River Severn at Shrewsbury, in the west of England. At the age of eight, he was sent to school in the town. In the same year, his mother died. In later life he recalled that long before she died, he had believed that people admired him for his perseverance and for his boldness; yet, at the same time, he was aware of feelings of vanity and self-contempt.[3] These features of Darwin's character were never to leave him.

As he grew, Darwin developed a passion for collecting anything from plants to coins. He also observed game and other wild birds. This was a great delight to him, and he later said, 'I was born a naturalist'. However, Darwin was no egghead. He fenced and competed at the high jump and at 'fives', a handball version of tennis played without racquets. Lessons bored him but he was fascinated by butterflies and other winged insects, and discovered a deep interest for natural history. He began to study birds and noted their habits and characteristics. At about the same time, he learned to shoot. He later recalled: 'I do not believe that anyone could have shown more zeal for the most holy cause than I did for shooting birds. This taste long continued and I became a very good shot.'[4]

His introduction to practical science came from his brother Erasmus, and Darwin developed a lifelong love of reading and a yearning to travel in remote countries. Unfortunately, his studies at Shrewsbury had never impressed anyone and his father, Dr Robert Darwin, decided that some changes were necessary for his son's own good. Dr Darwin was a formidable figure, weighing more than 24 stone (336 pounds), and Charles was in awe of him:

> His reverence for him was boundless and most touching. He would have wished to judge everything else in the world dispassionately, but anything his father had said was received with implicit faith.[5]

Darwin himself talked of his father's many admirable qualities, including his kindness and generosity. He respected his views on everything and believed he understood him: 'He was of an extremely sensitive nature, so that whatever annoyed or pained him did so to an extreme degree. He was also somewhat easily roused to anger.'[6]

However, Darwin's wife Emma gave an entirely different picture of her father-in-law when questioned by her son Francis in later life:

> Dr Darwin did not like [Charles] or understand him or sympathise with him as a boy. He was a fidgety man and the noise and untidiness of a boy were unpleasant to him... Everything in the household had to run in the

> master's [way] so that the inmates had not the sense of being free to do just what they liked. They never felt at ease, and used to be extremely glad when the Doctor went off on a long journey, and sorry to see him come back again.[7]

When Charles's older brother Erasmus left Cambridge and went to university in Scotland to complete his medical studies in 1825, Dr. Robert decided that Charles should go with him to study medicine. Charles was still only sixteen, but the decision was not questioned.

As an adolescent approaching adulthood, Charles was tall, slender, athletic and popular. His passions at Edinburgh University were not his medical studies but natural history, hunting foxes and shooting birds. Darwin recalled that this was not a direction that pleased his father, who railed at him: 'You care for nothing but shooting, dogs and rat-catching, and you will be a disgrace to yourself and all your family'. The outburst caused Darwin deep mortification, but he rationalised it to himself: 'My father, who was the kindest man I ever knew, and whose memory I love with all my heart, must have been angry and somewhat unjust when he used those words'.[8]

Towards the end of his life, reflecting on his relationship with his father, Darwin said: 'I think my father was a little unjust to me when I was young, but afterwards I am thankful to think I became a prime favourite with him.' However, Francis Darwin remembered his mother saying that the affection felt by his grandfather for his father sprang up only after the latter's return from the *Beagle* voyage.[9]

Darwin's study of medicine at Edinburgh amounted to just two seven-month academic sessions. He found some of the lectures 'exceedingly dull' and revealed his disgust at the idea of practical anatomy. He loved his lectures on chemistry and read avidly around a wide range of subjects, including science, mechanical arts, zoology, the system of classification of the animal kingdom, entomology, insects, shells and James Boswell's *Life of Samuel Johnson.* The list indicates his declining interest in medicine and an increasing fascination with the natural world.

He learned how to stuff birds and he studied French, but his waning interest in medicine did not go down well at home. His father made his views known in a letter from his sister Susan:

> I have a message from Papa to give you, which I'm afraid you won't like; he thinks your plan of picking and chusing [sic] what lectures you like to attend, not at all a good one; and as you cannot have enough information to know what may be of use to you, it is quite necessary for you to bear

> with a good deal of stupid and dry work: but if you do not discontinue your present indulgent [ways], your course of study will be utterly useless. Papa was sorry to hear that you thought of coming home before the course of lectures were finished, but hopes you will not do so.[10]

In his second year, Darwin decided to divide his study time between medicine and natural history, which took in geology. He must have been very disappointed. Looking back, he remembered the lectures as incredibly dull: 'The sole effect they produced on me was the determination never as long as I lived to read a book on geology, or in any way to study the science'.[11]

By the time he left Edinburgh, his interest lay solely in natural history. He was already familiar with the approved methods of collecting and identification and the value of careful observation and interpretation. His personal wish to avoid any connection with geology was not to be fulfilled. Darwin did not complete his course, but Edinburgh taught him a lesson about rivalry and jealousy in the world of science he was never to forget. He was warned off by Robert Grant, a zoology professor he had befriended, for discovering something in an area that Grant considered his own preserve. He was forced to come to terms with the idea of priority, by which an individual scientist is considered to have arrived at a certain idea first. Darwin told his daughter Henrietta about this incident. She later wrote: 'This made a deep impression on my father and he has always expressed the strongest contempt for all such little feelings – unworthy of searchers after truth'.[12]

When Darwin left Edinburgh without his degree, he no longer wanted to be a physician. In his autobiography, he remembered that his father had been 'properly vehement against my turning into an idle, sporting man, which then seemed my probable destination'.[13] Eight months later, Darwin went to study at Cambridge. His father's plan was that he should be ordained in order to pursue a future in the clergy. Darwin went along with the plan and looked forward to three agreeable years.

As might have been anticipated, Cambridge's effect on Darwin was determined more by the people he met and his driving interest in natural history than by any meaningful religious insights. Full of energy, he again spent much of his time shooting, fox-hunting and collecting beetles, and he soon discovered that a short period of intense study before each examination was all that was needed to achieve success. He left Cambridge in July 1831, fit, slender and just under six feet tall, with the unexpected reputation of being both an excellent shot and a first-class student. He had graduated with enough marks to be placed tenth out of 178 examinees.

His father's proposal that he enter the Church held little attraction, but in the absence of alternatives he agreed to go back to Cambridge in October, once the summer vacation had ended, to continue his studies with a view to becoming a member of the clergy. And then, with the arrival of one letter from Cambridge, everything in Charles Darwin's life changed. Before he had even graduated, he had entertained the idea of visiting the Canary Islands off the coast of northwest Africa, and in particular the island of Tenerife. He had dreamed of living in the tropics, observing the scenery and vegetation. He had even studied practical geology in the mountains of Wales in preparation for the geological formations of the Canaries. Then, one day, he learned from his former tutor at Cambridge that a British Navy surveying ship leaving England in the autumn needed the services of a suitable young man interested in science and natural history to be a companion to the captain. The journey would take at least two years. Only a gentleman naturalist suitably recommended need apply.

Three months later, HMS *Beagle* left the Royal Navy base at Devonport, within the harbour of Plymouth on the southwest coast of England, with Charles Darwin on board. For the next five years, life was one long adventure as the *Beagle* visited Brazil, Argentina, Chile, Tierra del Fuego, Peru, the Galapagos Islands, Tahiti, New Zealand, Australia and a score of islands along the way. For Darwin, it was an opportunity to indulge his fascination for collecting, hunting and shooting. As the *Beagle* slowly surveyed the coastline of South America, Darwin collected rare species. He stuffed and prepared animals and birds for dispatch back to England, and recorded his impressions of scenery, people, customs and strange sights. He began the voyage little more than a youth and returned an independent and seasoned traveller.

The experience affirmed his interest in geology, the subject that had bored him so profoundly when he was at Edinburgh. It also led him to Lyell, an important patron and mentor who would prove important to Darwin twenty years later. Finally, it led to his acceptance by some of the most eminent scientists of his day, fascinated by the accounts of his adventures and the collections he had sent back to England.

When the *Beagle* docked at Falmouth nearly five years after she had set sail, Charles Darwin stepped ashore with no doubt about the reception his family were planning for him, because some months before he had received a letter from his sister Susan. The letter referred admiringly to Darwin's 'fame and glory' (the result of his letters being read out at Geological Society meetings), and the changed opinions of him in their household:

> Papa and we often cogitate over the fire what you will do when you return as I fear there are but small hopes of you still going into the Church. I think you must turn Professor at Cambridge.[14]

In fact, Darwin was too busy to think about a career. Given a generous allowance by his father, he found scientists in various fields to write proper scientific descriptions of all the zoological specimens he had dispatched or brought back with him. However, his interests during the voyage had been mineral rather than animal (which must have surprised those who knew how much he had suffered with his geology studies at Edinburgh) and he kept for himself the geological specimens.

> All my affairs indeed are most prosperous; I find there are plenty, who will undertake the description of whole tribes of animals, of which I know nothing... I hope to set to work, tooth and nail, at the Geology, which I shall publish by itself.[15]

By mid-December, Darwin had settled in lodgings in Cambridge. He began work on his geological collections and was writing an account of his *Beagle* diary for official publication by the government department that administered the Royal Navy. Then, a few months later, an appearance before geologists in London led to a friendship that was to help ensure his place in history.

In February 1837, Sir Charles Lyell, by then the pre-eminent geologist in England and President of the Geological Society, referred in his presidential address to the quality of Darwin's geological ideas in the letters he had sent back from the *Beagle* voyage. If Darwin was pleased by such praise, his father was ecstatic. His sister Caroline told him, 'My father is extremely pleased by Mr Lyell's friendship for you', adding that Dr Robert was begging for a letter from his son giving as much detail as possible of the relationship between the two men.[16]

What had so impressed Lyell was Darwin's descriptions of the physical changes wrought on the coastline of Chile by an earthquake at Concepción in January 1835. When Darwin and Robert FitzRoy, the captain of the *Beagle*, visited the area, they both realised that the effect of the earthquake had been to elevate an extensive section of the coastline. Lyell was excited because this indicated that he was right in his beliefs that similar seismic action over immense stretches of time would account for changes in the surface of the earth. News of the earthquake had reached England by the summer of 1835, and a few months later Lyell, searching for evidence to support his position, wrote to his main critic: 'Give me but a few thousand centuries, and I will get contorted and fractured beds above water in Chili, horizontal ones in Sweden, etc'.[17]

Within a month of returning, Darwin was invited to dine with Lyell at his house in the fashionable Bloomsbury area of London. A week later, he told a friend that he had seen Lyell several times and that the latter had been extremely friendly and kind. 'You cannot imagine how good-naturedly he entered into all my plans', he wrote.[18] A few months later, he indicated his own reactions:

> I have read some short papers to the Geological Society, and they were favourably received by the great guns, and this gives me much confidence, and I hope not a very great deal of vanity, though I confess I feel too often like a peacock admiring his tail. I never expected that my geology would ever have been worth the consideration of such men, as Lyell, who has been to me, since my return, a most active friend.[19]

At that point, Darwin could not have realised how important his friendship with Sir Charles Lyell would become nor did he realise how important to Lyell's professional standing was the geological evidence he had brought back from the voyage.

Lyell had initially had a hard time convincing critics of the truth of his Huttonite theory of a world formed over limitless time. He was aware that his ideas were viewed as heretical because they directly contradicted the information presented in the Bible. Lyell's luck began to change in March 1838 when Darwin presented a paper to the Geological Society (to which he had been elected almost immediately after his return). The paper was entitled 'Volcanic phenomena in South America' and argued strongly in support of Lyell's idea that changes to the surface of the world were the result of gradual changes caused by seismic events.[20]

Darwin's eyewitness account of the reshaped Chilean coastline helped to change the scientific perception of Lyell's work. Lyell noted afterwards:

> I was much struck with the different tone in which my gradual causes was treated by all ... from that which they experienced four years ago, [when they had been treated] with as much ridicule as was consistent with politeness in my presence.[21]

Darwin won the everlasting gratitude of Lyell, who from that moment took on the role of the young man's mentor. It was to be more than twenty years before Lyell found an appropriate way to repay him and the subject area would be the much debated species question.

Darwin had already opened his first notebook on the species question in the summer of 1837. In that and succeeding books and manuscripts, he wrote down fragments of his ideas on science, on methodology, and about when it might be proper to generalise or speculate about causes. In 1838, he admitted his fascination with the subject to Lyell:

> I have lately been sadly tempted to be idle, that is, as far as pure geology is concerned, by the delightful number of new views, which have been coming in thickly and steadily, on the classification and affinities and instincts of animals – bearing on the question of species. Notebook after notebook has been filled with facts which begin to group themselves *clearly* under sub-laws.[22]

Lyell would have been fascinated to know what Darwin was thinking. In fact, Darwin had been influenced by Lyell's ideas, but unlike Lyell, he had decided to adopt the scientific methodology of Francis Bacon: observe first and only then design a theory to fit the observations. He decided to collect as many facts as he could on the subject of what was then called 'transmutation' in order to establish one way or the other whether or not species could mutate into other species.[23]

During the 1840s, Darwin's own domestic life was slowly evolving. He had married his first cousin Emma Wedgwood in 1839 and immediately set about creating a large family – they were to have ten children, three of whom did not survive childhood. In 1842, he bought Down House, a former farmhouse on the outskirts of London, and began extending it to accommodate his expanding family. Down House proved to be an ideal setting for Darwin to mull over his research and ponder where it might lead him. His characteristically unhurried pace was made possible by the fact he was already financially secure, due to an annual income of several thousand pounds from investments made on behalf of his wife when they married. When Darwin's father died in 1848, his personal financial well-being was further transformed with an inheritance that generated an additional income of £3,000 a year. Although Darwin's financial security was guaranteed during this period, his personal life was overshadowed by his own ill-health. Following the *Beagle* expedition, Darwin had begun to suffer acute bouts of illness and it is possible that he picked up an infection such as Chagas disease, transmitted through the bite of an infected bug, on the expedition itself. Whatever the cause, for months at a time Darwin was unable to work, incapacitated by stomach pains, sickness and heart palpitations.

By 1844, seven years after writing down his first thoughts about transmutation, Darwin had filled many notebooks and manuscripts with his ideas. He had collected every relevant fact he could find and written a 230-page essay entitled 'Natural Selection'. Within weeks, however, he instructed his wife that it should be published only in the event of his death. We do not know his reasons for this decision: some have argued that he did it out of respect for his wife's religious feelings, others that he was concerned about the civil unrest that his ideas might provoke.

In 1845, he began writing the new version of his *Beagle* journal, which was to prove such a success later that year. He then turned away from his *Beagle* adventures and the mystery of transmutation, and began a scientific study of barnacles that was to last for eight years.

When, in 1854, he eventually looked up from his completed study of barnacles, his mind had once again turned to the problem of how new species arrive in the world. Exactly five years later, he sent *On the Origin of Species* off for publication, and saw it burst upon an unsuspecting public still in thrall to the Christian faith.

CHAPTER 4

Emperor's clothes

ON THE centenary anniversary of the publication of *On the Origin of Species* in 1959, scientists around the world sought appropriate ways to celebrate the book in which Darwin had presented his 'discovery' of the theory of evolution. For the most part, the activities were predictable: meetings were organised, new research instigated, and papers and books of quality written to mark Darwin's brilliant insight.

One man, however, working barely a hundred miles from where Darwin had written of his great discovery, had stumbled upon a huge problem. Rather like the wicked stepmother at Snow White's christening party, he chose to voice his concerns just before the centenary celebrations.[1] Cyril Dean Darlington, Sherardian Professor of Botany at Oxford University, had a question: by what thought process had Charles Darwin actually arrived at his ideas about evolution?[2]

Darlington pointed out that he could not find, in all the accounts of Darwin's work published up to that time, any suggestion that some original germ in Darwin's mind had led inexorably to the full development and enunciation of this big idea.

For non-specialists, Darlington summarised the main points of Darwin's *Origin* as follows. When bred, all kinds of things vary in character. When humans breed animals they can, by careful selection, produce new varieties or even new species. Nature, over a vast period of time, does the same with all animals and plants. Nature's selection depends in part on the removal of those least able to survive. In moving to new regions and occupying different habitats, animals and plants adapt to diverse conditions. By this process of adaptation, existing species are transformed into new species.[3]

This process, declared Darlington, explains common descent, similarities of structure and the development of living organisms. It also explains other things: why fossil forms within stratified rocks can be seen to change over time; why some species' characteristics differ between the 'old' and 'new' worlds; and why there can be a great diversity of birds or molluscs in a chain of small islands, but much less diversity in a large continent. This theory meant that the habit of cross-breeding, the sterility of hybrids and dozens of other problems could be usefully discussed in a way that was not possible when people believed that species had been individually created by an inscrutable supernatural power.[4]

However, despite his enthusiasm for the theory, Darlington was not happy. He pointed out that the *Origin* did not contain any account of how Darwin had come by his ideas.[5] The opening words of the *Origin* suggest that the *Beagle* voyage marked the beginning of the debate around the species question, whereas in fact natural historians had been discussing the problem for decades, if not centuries. Darwin referred constantly to 'my theory', but Darlington asked: 'What exactly were Darwin's ideas, his own distinctive ideas, and what do we mean by Darwinism?',[6] before accusing Darwin of simply collecting 'the evidence of Lyell and his great friend, the botanist Joseph Hooker, and editing it as material for his own arguments'.[7] He also argued, while mentioning no names, that 'Darwin's unawareness of what his contemporaries were thinking matched his unawareness of what his predecessors had written.'[8]

On the occasion of the centenary anniversary of *On the Origin of Species*, Darlington felt that he was entitled to ask some rather direct questions that went quite against the grain of conformist academic praise for Charles Darwin. He decided to be extremely direct: much more direct than scholars had been before, and more direct than most scholars have been since:

> How is it, we may now ask ourselves, that so much obscurity overhangs the development of the greatest of modern ideas? After a hundred years we are almost as uncertain of the authorship or editorship of Darwin's writings as we are of those attributed to Homer or Hippocrates. This is due, on the one hand, to the fact that people who investigate the history of science are historians who are not entirely clear about the meaning of its ideas. They also often believe what the discoverer writes about his own discoveries, which, as we see, is not a wise thing to do. On the other hand, among scientists there is a natural feeling that one of the greatest of our figures should not be dissected, at least by one of us. The myth should be respected.[9]

In fact, prior to 1959, accounts of Darwin's life and work had been very much a family industry, and a limited one at that. Following Darwin's death in 1882, a very brief autobiography, which had been written towards the end of his life together with selected correspondence with family, friends and academic colleagues, was collated and edited by his son Francis, and published in 1887.[10]

Nearly twenty years later, Darwin's first attempt at setting out a theory to solve the species question was discovered in a cupboard in Down House. Darwin had written a thirty-page sketch of the theory in 1842 and the new discovery was an expanded version of this. It ran to two hundred and thirty pages, and had been completed in 1844. This was the manuscript Darwin had instructed his wife to have published only in the event of his death. It had not been published, and had been read in its entirety only by his greatest friend, the botanist Joseph Hooker. In 1909, Francis Darwin published both sketch and essay as the *Foundations of the Origin of Species*.[11]

In his autobiography, Darwin never revealed what had led him to recognise the forces of divergence and modification, which lay at the heart of his world-changing book. He claimed no great insights from his time on the *Beagle*, except that after his return he had never stopped collecting facts 'bearing on the origin of species; and I could sometimes do this when I could do nothing else from illness'. He said that in October 1838 he had 'chanced to read' the political philosopher Thomas Malthus's *Essay on Population*,[12] and it was in that book that he had found the key to the idea of natural selection in the animal world. He went on:

> Being well prepared to appreciate the struggle for existence which everywhere goes on, it at once struck me that under these circumstances, favourable variations would tend to be preserved and unfavourable ones to be destroyed. The result of this would be the formation of new species. Here then I had at last got a theory by which to work...[13]

However, one American – writing at the same time as Darlington and also with his eye on the centenary celebrations – was not convinced.[14] For some time, the poet, anthropologist and science writer Loren Eiseley had been reading copies, some more than a hundred years old, of *The Annals and Magazine of Natural History*, which had been Darwin's favourite scientific journal. The question that bothered him was simple, but the answer held profound consequences for Darwin's reputation. If Charles Darwin did not chance upon Malthus (and form his ideas about natural selection) until October 1838, how was it that he was making notes on that very idea in the early months of 1837, shortly after opening his first notebook on transmutation?

It was a very pertinent question because those old copies of the *Annals* suggested that Darwin had relied for his inspiration not on the celebrated ideas of Thomas Malthus, but rather on an impecunious young naturalist, unknown and uncelebrated, who was living at the time in a small town just outside London.[15]

CHAPTER 5

The chemist from Tooting

FOR Loren Eiseley, the clinching piece of evidence was the word 'inosculate'. He was sure he could construct a case from many other pieces of evidence he had found, but that one word – meaning 'to unite or be united so as to be continuous; to blend; to intertwine' – convinced him that Darwin's claim that Thomas Malthus had been his inspiration for the idea of natural selection was absolutely untrue.[1] He was similarly convinced that Darwin had taken that central idea from the work of a colleague, without sufficient acknowledgement.

Eiseley, who was Benjamin Franklin Professor of Anthropology and the History of Science at the University of Pennsylvania and a lifelong supporter of Darwin and his ideas, knew he was about to upset many friends and colleagues. However, he felt that Edward Blyth, a pharmacist from the village of Tooting (now a suburb of London), should at last be given credit for the idea that Darwin had exploited.

In Eiseley's view, selection was a simple concept. There were no intricate propositions, no complicated mathematics. In simpler forms, the idea of selection had been known to breeders for thousands of years, and the relationship between selection and the struggle for existence had been glimpsed by natural philosophers a hundred years before Darwin (and, significantly, by Charles Lyell himself). However, its creative aspect beyond the bounds of species had not been grasped, because of religious prejudice and because the sheer length of geological time inherent in understanding the idea had remained largely unappreciated in the first half of the nineteenth century.

Eiseley was well aware of Darwin's claim that he had come by his theory of natural selection as a result of reading Malthus's *Essay on Population* in 1838, two years after returning from his voyage on the *Beagle.* Eiseley felt that this was an account 'hallowed by tradition', which by the late 1950s had taken on the appearance of a fact. After reading those long-forgotten magazine articles, however, Eiseley became convinced that

Darwin had taken the idea of natural selection – although not the phrase itself – from ideas that Edward Blyth had written about between 1835 and 1837.[2]

From his research, Eiseley believed that the period between Darwin arriving home from the *Beagle* voyage in October 1836 and the opening of his first notebook on the species question in 1837 was crucial. Despite having read Lyell's *Principles* on the voyage, Darwin was still struggling with the problem of the rapidity of organic change. He still believed the history of the earth could be measured in thousands of years, rather than millions or even thousands of millions, despite Lyell's evidence. He thought that if species did change from one form into another, then the process must be one of sudden change; there would simply have been no time for gradual change. Darwin actually imagined that an organism would jump from one form into another (*per saltum*, 'at one bound') to allow it to cope with the new environment in which it found itself.[3] An organism that took too long to adapt would perish in alien surroundings. An organism, and therefore a species, could survive only if it changed instantaneously into a different form.

According to Eiseley, this view of how species originated, which Darwin held in 1837, was entirely different from the process he was to describe when he wrote the *Origin* 18 years later.[4] He also knew that his fellow academics had turned a blind eye to this fundamental change:

> The fact is that an important shift in Darwin's thinking remains undocumented. It has not even been discussed. In the *Origin*, slow and imperceptible transformations extending over vast ages of time have replaced his early and immature speculations on organic change *per saltum*.[5]

What had prompted this change? Eiseley wrote: 'the published notes fall silent'.[6] He turned to Darwin's sketch of 1842 and the unpublished essay of 1844.[7] He felt that since they dated from the earlier years of Darwin's work on evolution, they might yield clues to the change in his thinking. As he followed this path, Eiseley stumbled across Edward Blyth, and the part he had played in changing Darwin's ideas.

Blyth, a year younger than Darwin, did not come from a rich family and had attended a trade school rather than university. An avid reader who was fascinated by natural history, he used a small legacy to buy a chemist's shop in Tooting, but his business failed to flourish; he spent more time with the books and natural history collections at the British Museum than with his customers.[8] Between 1835 and 1837, Blyth wrote three articles on the species question for *The Annals and Magazine of Natural History*. Darwin's associates, including Lyell, had all had articles published in its pages, and Eiseley was convinced that it was illogical in the extreme to assume that Darwin could have been unaware of Blyth's articles. In fact, so keen had Darwin been to keep up with the latest ideas in natural philosophy that he had arranged for the *Annals* to be delivered to ports visited by the *Beagle*.

Eiseley also discovered that Darwin had held in his hands and made use of, for scholarly purposes, the 1835 issue of the *Annals* that contained Edward Blyth's first essay on species and the idea of natural selection. It is likely that Darwin read it on the journey home to England having received it in his mail on arrival in Australia some time after its publication.[9] Moreover, he discovered that the volumes had been scanned with great care. Yet when he examined Darwin's transmutation notebooks, he could find no reference to Blyth himself.

In *On the Origin of Species*, Darwin wrote that he valued Blyth's opinion on a considerable range of topics 'more than that of almost anyone',[10] but he never made a direct reference to his awareness of Blyth's ideas in the *Annals* articles. 'One begins to get the feeling that something more than chance is at work in this situation', Eiseley wrote, as he pointed out that Darwin opened his first notebook on transmutation in 1837, only weeks after Blyth's final paper commenting upon the idea of natural selection was published in the *Annals*.

Eiseley wanted to find, in either the trial essays of the 1840s or the *Origin* itself in 1859, some direct proof that Blyth's ideas had influenced Darwin.[11] In the end it was the trial essays that provided the information he sought. Without them, Eiseley admitted, it was unlikely that 'the dim outlines of the carefully hidden trail would ever have been perceived'. The word 'inosculate', 'which never has had a wide circulation and which was not to be found in Darwin's vocabulary before this time', was the key piece of evidence for his contention that Darwin had taken the idea of natural selection from Blyth without attribution.[12]

In 1835, at the age of 25, Edward Blyth was familiar with Charles Lyell's idea that much extinction could be accounted for by the struggle for existence that had been going on for millions of years. In that struggle, the most fertile would always prevail over the most sterile.[13] He was also aware of Lyell's belief that living forms revealed considerable variation across geological time, but that such variation was designed by the Creator to ensure the survival of new forms placed in changed physical environments. In those parts of the world that remained unchanged by geological forces, all species remained perfectly adapted to their environment and therefore variation could not occur.

However, Blyth, who was also a Christian, saw the world differently to Lyell. He saw a world held in a tight dynamic balance by natural forces that kept species inside their own environment. He believed that if they migrated, they could not survive.[14] Within these environments, he believed, species certainly faced a struggle to survive, but they were perfectly adapted and able to deal with any external threats, as well as any variation threatening to lead to a change in species.

In his articles in the *Annals*, Blyth suggested that in each species there seemed to be a tendency to some particular kind of variation. This tendency would only be counteracted, he suggested, 'by the various crossings which in a state of nature must take place and by the … law which causes each race to be chiefly propagated by the most typical and perfect individuals'.[15]

The fact that the original form of a species was unquestionably better adapted to its natural habitats meant that no new modification of that form could survive in nature. Such adaptation, he argued, was the Creator's way of eliminating any animal unfit for its environment. Blyth, however, did not view this tendency to vary as an invisible force for advancing the survival of the fittest.

Here, Eiseley argued, Blyth had come very close to spelling out the idea that through adaptation, extended change was possible but because of his religious beliefs he did not make that intellectual leap. Rather, he dismissed extended change as a possibility and drew back from the principle at the heart of the theory of evolution.

Eiseley obviously sympathised with Blyth at this point of his account of the young man's work, for where Blyth drew back, Darwin forged ahead and used Blyth's idea in a positive form to advance the idea of change rather than submit to its restrictions.

As a researcher, however, Eiseley had gradually become much more interested in discovering exactly how Blyth's ideas had affected Darwin's understanding of the process of natural selection. If he could show that by 1837 they had had a significant effect on Darwin, then Darwin's claim that his theory of natural selection had been inspired by reading Malthus in October 1838 would be shown to be untrue.

Eiseley began with Darwin's ideas about species 'jumping' from one form to another. In his species notebook, just a few months after Blyth's final article had appeared in the *Annals* in 1837, Darwin considered the possibility of distinct species uniting or blending together. He wrote: 'If distinct species inosculate so we must believe ancient ones [did too]…

[therefore] not gradual change or degeneration from circumstances'. The word 'inosculate' had appeared in Blyth's article and this seems to have been the only time Darwin ever used it.

This use of 'inosculate' convinced Eiseley that it was Blyth and not Malthus who had really influenced Darwin in his earliest ideas on the species question. As extra evidence, he pointed to the fact that even Darwin's own son, Francis, had expressed surprise that his father should have had to turn to Malthus for inspiration in 1838, when his transmutation notebooks indicated clearly that he had already given vent to the essential aspects of the principle and had discussed the ideas of both natural and sexual selection in those notebooks one whole year before stumbling upon Malthus.[16]

The difference between Blyth's approach and Darwin's was affected by their religious and theoretical beliefs. Blyth was constrained by the ideas of Lyell in terms of perfect adaptation. Darwin, however, while also believing in the principle of perfect adaptation in settled environments, believed that new environments created by geological upheaval would be populated by species that migrated there from the nearest land. In such circumstances variation and change were possible, because these migrant species would no longer be perfectly adapted to this new environment, and yet they would have to become perfectly adapted if they were to survive.

Eiseley believed his comparison between the writings of the two men at this time revealed an amazing resemblance between Blyth's thoughts and those written out by Darwin in his notebooks. In his 1835 paper, Blyth had argued that at the root of any logical system of classifying animals was what he called 'a law of irregular and indefinite radiation' (the notion that species are somehow programmed to modify away from ancestral forms), and that the ways each successive type had been modified were always in direct relation to particular localities or to peculiar modes of procuring sustenance. Then, having earlier denied that these forces could lead to indefinite divergence, Blyth theorised that 'just as man is able to affect the physical constitution and adaptations of domestic animals, so wild nature might achieve the same success'.[17]

For Eiseley, this was an amazing and contradictory passage. Had Blyth somehow been able to escape the deeply embedded religious beliefs of his own age, which insisted that all living creatures were the individual creations of God? Whatever the answer, Eiseley believed this passage indicated that Blyth had come 'within a short, usable compass' of spelling out a remarkably complete abstract of the theory of evolution. He also discovered that all the leading principles of Darwin's theories – the struggle for existence, variation, natural selection and sexual selection – were fully expressed in Blyth's 1835 paper.[18] Eiseley outlined one section particularly:

> It is a general law of nature for all creatures to propagate themselves: and this extends even to the most trivial minutiae, to the slightest peculiarities; and, thus, among ourselves, we see a family likeness transmitted from generation to generation. When two animals are matched together, each remarkable for a certain peculiarity, no matter how trivial, there is also a decided tendency in nature for that peculiarity to increase; and if the produce of these animals be set apart, and only those in which the same peculiarity is most apparent, be selected to breed from, the next generation will possess it in a still more remarkable degree; and so on, till at length [from] the variety I designate a breed is formed, which may be very unlike the original type.

Blyth was aware of Lyell's belief that when geological forces caused environmental circumstances to change, a species must perish with its locality, and argued that just as the surface of the earth varies, so do its productions and inhabitants. Eiseley felt it was beyond doubt not only that Darwin's thinking was influenced by Blyth's articles, but that many pages of Darwin's preliminary essays and of the *Origin* itself are spent in addressing Blyth's belief in what he termed a 'species barrier', where all extinct past forms are without representation in the present. Because of this influence, Eiseley believed, Blyth deserved his place in the family tree of *On the Origin of Species*:[19] 'He is not an isolated accident [but] one of the forgotten parents of a great classic'. Eiseley wondered how to trace exactly how Darwin had incorporated Blyth's ideas into his own work: 'Darwin's shadow, grown to almost superhuman proportions, lies massive and dark across the early portion of the century. How can one find, even in this similarity of ideas, more than the accidental repetition of like thoughts by different men?' And yet, with patience and forensic precision, Eiseley teased out the damning detail.

Eiseley pointed out other echoes of Blyth's articles in Darwin's handwritten theories; examples where he felt pure chance 'was so remote as to be almost nonexistent'.[20] Blyth refers in one paragraph to Ancon sheep and then, a little later, to some odd mutations including 'donkey-footed swine, tail-less cats, back-feathered, five-toed and rump-less fowls together with many sorts of dogs.' In his essay of 1844, Darwin, discussing 'sports' or hereditary monsters, duplicates the list in almost the same order, and mentions 'Ancon sheep, rump-less fowls and tail-less cats'.

Discussing food and its effects on animals, Blyth refers to the fact that 'herbivorous quadrupeds which browse the scanty vegetation on mountains are invariably much smaller than their brethren which crop the luxuriant produce of the plains'. In his essay, Darwin holds that 'external conditions will doubtless influence and modify the results of the most careful selection; it has been found impossible to prevent certain breeds of cattle from degenerating on mountain pastures'. Where Blyth discussed 'hybridity, dominance and the re-emergence of suppressed characteristics in the third generation of a breed of animal', Darwin, again in the essay, expressed similar views.

Blyth observed that domestic animals supplied with an abundance of food become bulky and lazy with powerless or underdeveloped muscles; Darwin devoted a whole section to this notion, came to similar conclusions and expressed them in both the essay and, eventually, the *Origin.*

Blyth discussed protective colouration as a device in the struggle for existence, and used the metaphor of grouse as 'brown heather'; Darwin used the same metaphor, picturing the red grouse 'the colour of heather'. Blyth spoke of the ptarmigan as 'snow in winter'; Darwin wrote of the alpine ptarmigan as being 'white in winter'. Where Blyth, talking of a falcon in relation to its prey, dwelled upon the bird's 'great powers of sight', Darwin mentioned hawks as 'guided by eyesight to their prey'. Darwin echoed Blyth when he described the pruning effect exercised by birds of prey, which helps to keep the cryptic colouration of small mammals and ground-dwelling birds, like the grouse, uniform and constant. When Blyth discussed the homing instinct and gave as an example the Australian aboriginal as being subject to this 'intuitive impulse', Darwin used both the idea and the Australian 'savage' as an example. Finally, Blyth's contention that the instinctive shamming of death is 'a characteristic of certain animals' turns up briefly but critically in both Darwin's sketch and essay.

Eiseley believed, even making some allowance for the accidental use of the same sources, that the effect of his research was cumulative. He argued that these many similarities could not be explained by chance and that Darwin had plundered Blyth's articles for the ideas which underpinned the thinking that led to *On the Origin of Species.*

Eiseley also found worrying indications of the use of unacknowledged material from another source. In the essay of 1844, Darwin had written:

> In the case of forest trees raised in nurseries, which vary more than the same trees do in their aboriginal forests, the cause would seem to lie in their not having to struggle against other trees and weeds, which in their natural state doubtless would limit the conditions of their existence.

In his book *On Naval Timber and Arboriculture*, written thirteen years earlier, Patrick Matthew had fully anticipated the idea of natural selection and had even used an almost identical phrase. In addition, he had written the following passage:

> Man's interference, by preventing this natural process of selection among plants, independent of the wider range of circumstances to which he introduces them, has increased the difference in varieties particularly in the more domesticated kinds.[21]

Eiseley claimed the content of the two passages suggested that by 1844, Darwin was well aware of Matthew, and that the use of both principle and phraseology in the essay makes it less easy to accept Darwin's ingenuous claim, when faced with Matthew's accusation of plagiarism after the publication of *On the Origin of Species*, that 'one may be excused in not having discovered the fact [natural selection] in a work on Naval Timber'.

By now Eiseley, really intrigued, began to wonder, as had others before him, where Darwin might have come across the phrase 'natural selection' if he had not read Matthew. He knew that Blyth had never used the expression, and yet it was so distinctive that Darwin had even used it as a title for his 1844 essay. The nearest thing Eiseley could find to Darwin's first use of the phrase occurred in Matthew's explanation of why trees in nurseries seem to vary more than those in the wild. Matthew speaks of 'this natural process of selection' in 1831. Darwin, in turn, used the expression 'natural means of selection' in both the sketch of 1842 and the essay of 1844, before dropping the words 'means' and 'of' and referring only to 'natural selection' when using the phrase later in that work. Eiseley was convinced that Darwin had taken that, too, from Matthew, and claimed the reworked paragraph proved that Darwin had read Matthew before writing both the sketch and the essay.

As a result of his research, Eiseley, who had long championed Darwin, claimed that although the latter had often been depicted as a simple, forthright man, he had in reality been an enormously complex human being. 'It would seem', he wrote, 'there was a genuine and understandable hunger to possess the theory as totally his own. One can only conclude that Darwin was solitary and elusive beyond even what his family has recorded'.[22] Eiseley went further, and echoed Darlington in criticising his own colleagues for their lack of vigour.

> There have been also the complications introduced through the unconscious process of myth-making, the desire, in other words, to keep this

> man and his discovery inviolate – a unique act of genius without precedent and without precursive steps. There has thus arisen a tendency to see Darwin's forerunners as having no relation to his own accomplishment. They are dismissed, as Darwin was inclined to dismiss his own grandfather as 'part of the history of error', as speculative, as lacking in facts.

Yet Darwin, having relied on Blyth for many of the ideas in his notebooks, now had a big problem. If he published the ideas contained in the sketch and the essay, he would take the chance of being exposed as a plagiarist. Moreover, a great deal of Darwin's thinking in general on the species question was derived from the ideas of Charles Lyell and Augustin de Candolle. Eiseley believed that Darwin could not have named Lyell directly as the source of his ideas because then he would have been quoting the ideas of a deeply religious man who was publicly opposed to evolution. Thomas Malthus was a different proposition. Active in quite a different field, and the source of much of the thinking on the struggle for existence in early nineteenth-century Britain, it was convenient for Darwin to have recourse to Malthus.

> There are, it is true, a few references to Malthus in the trial essays before the *Origin*, but not to the exclusion of other writers such as de Candolle. If one turns back to Darwin's letters of the 1840s, one gets the same impression of neglect. Though Darwin often wrote to Hooker, Gray and Jennyns about his work and about the struggle for existence, Malthus remains unmentioned.[23]

In fact, in September 1857, Darwin was to claim as his greatest influences de Candolle, Charles Lyell and the Reverend W. Herbert (a long-time correspondent of Darwin's). There was no mention of Malthus. Eiseley quoted the research of a colleague, Gerald Henderson,[24] to show that only twenty years after opening his notebooks on species did the name of Thomas Malthus 'bulk larger' in Darwin's public declarations about the origins of his views on species.

In the light of these difficulties of acknowledging his precursors, Darwin decided to do nothing. His 'hesitations, long delays over publishing and almost neurotic anxiety' could now be better understood. 'He had his secrets and ... he had his justification for them', Eiseley concluded.[25] In 1844, when he had finished writing out the 230 pages of the essay, Darwin gave it to his wife and asked her to organise its publication in the event of his death.[26]

Generations of biographers and supporters have alluded to the essay as a great work that was kept hidden for twenty years, but in fact it was an embarrassment for Charles Darwin. Had it been published, his peers might have recognised Blyth's influence, in addition to the influence of both Lyell and de Candolle. Darwin might then have become a laughing stock in the natural history world he dreamed of conquering. Instead, Darwin was at work on a plan to get around the problem of giving credit to those whose ideas he had passed off as his own. It would allow him neatly to reveal to the world that most of his ideas on the species question had come to him, and been recorded in his journals, years before he had ever heard of Edward Blyth or opened his first species notebook.

CHAPTER 6

The sting

AS Loren Eiseley was coming to terms with Darwin's treatment of Blyth in the summer of 1957, another academic was slowly working his way through the scientific notebooks Darwin had completed during the *Beagle* voyage. Howard Gruber, a cognitive psychologist, was searching among thousands of pages of Darwin's notes for any early expression of belief in or concern for the idea of organic evolution while he was on the voyage.[1] He had the idea of tracing Darwin's intellectual development, through the study of his notebooks, during the two years immediately following his return from the expedition.[2] However, try as he might, Gruber could find almost no hint of evolutionary thought in the scientific notes or letters Darwin had written while on the *Beagle*.[3]

He could not avoid the conclusion that during the years Charles Darwin had spent on the voyage, he was preoccupied with geological questions rather than the origin of species. In fact, he could find no evidence in the *Beagle* papers, despite all Darwin's experiences aboard ship, that he had expressed interest in any scientific discipline other than that of geology.

Gruber was, to say the very least, surprised by this discovery. After all, Darwin's fame before the *Origin* was based on his Galapagos experiences while on the *Beagle*. Gruber now had a mystery to solve. If the world connected Darwin's theory of evolution with what he had seen on his visit to the Galapagos Islands, how, in the absence of any hint of evolutionary ideas in the journal, could such a connection have ever been made?

As he worked on how Darwin's theory had developed before and during the *Beagle* years, Gruber discovered that Darwin's early thinking about the species question had been disappointingly traditional. It had begun with Lyell's idea of a stable, harmonious system on earth in which all organic beings were perfectly adapted to each other and to their physical environment in a fashion ordained by the Creator. Then, as he had begun to accept Lyell's modern belief in the constantly changing geological profile of the earth's surface, a contradiction within his point of view had developed as he tried to reconcile unchanging and perfectly adapted species with a world undergoing constant and violent geological change.

If Darwin had ever dwelt on this contradiction, Gruber felt, it was only during the final months of the voyage. There was little doubt that when Darwin got home, he still leaned more towards the idea of a Creator than that of organic evolution. His *Beagle* notebooks proved it. It was to be a further twelve months before Darwin opened the first of his notebooks on the subject of species transmutation.[4] Gruber, disappointed with his lack of success (and wrestling with Darwin's handwriting), suddenly had a brilliant idea. He knew that Darwin's views on the species question had changed radically between the end of the *Beagle* voyage in October 1836 and 1844. He decided that one way of documenting how Darwin's views on species had developed would be to make a detailed comparison of the two accounts Darwin had written of his *Beagle* voyage. The first account had been published in 1839 and the second in 1845.

What he discovered in his comparative study threw a very revealing light on how Darwin had exploited what he had actually observed on the Galapagos Islands, and what he had learned about the islands and the species that exist there from others in the years after he arrived back in England.

After painstaking comparison, Gruber was surprised at the 'many and sometimes subtle' differences he found between the way events had been recorded in the 1839 account of the journey, and how the same events had been recorded in the second account, published six years later, which had been read avidly all round the world.[5] Throughout the second account, Darwin had inserted paragraphs dealing with evolutionary ideas that could only have been written by Darwin the evolutionist, and not by Darwin the geologist, as he was on the *Beagle* voyage. Gruber commented:

> Taken out of their hiding places and strung together, they form an essay which gives almost the whole of his thought. He used two methods of concealment: fragmentation and dispersal of the relevant passages, a paragraph here and there throughout the book; and omission of one vital ingredient, the principle of natural selection acting to produce new species.[6]

No one who bought either the 1839 or the 1845 editions of the *Beagle* journal had the opportunity to check them against the original journals from the voyage, which were still held in Darwin's possession, so it would have been accepted without question that the views and ideas expressed in the later account were those Darwin had held while writing up his entries on the ship.

Gruber was convinced that Darwin had interpolated many important themes relating to the species question into his account, and made them visible enough to be picked up by any reader who happened to be interested. Although Gruber felt that neither version could be read as a full and coherent account of a contemporaneous theory, he felt the differences between them reflected changes in Darwin's theoretical mindset over the intervening years.

Gruber saw the changes Darwin had introduced in 1845 as 'a flashing arrow to the way in which he had managed a great secret – by telling much of it in print, in this fragmented way, holding back only a few crucial points'. He considered the implications: 'It seemed to me that a person untrained in historical research, like me, ought not to have made this little discovery about Darwin. By 1956 it should have been well known.'[7] He added that planting fragments of his theory in an apparently surreptitious way must have been motivated by a desire to establish his claim to priority, like an explorer who cannot actually colonise a territory but instead only plants the flag. This would have been of primary importance to Charles Darwin. These extracts would enable Darwin to claim that he had been aware of certain ideas long before those he might otherwise have had to acknowledge as forerunners.

Darwin's tactics seemed to have worked. Of those who bought or read the 1845 version of the journal, very few would have referred back to the 1839 edition. The 1845 edition was accepted by new readers and the generations that followed as representing Darwin's thoughts and ideas while on the voyage. To all intents and purposes, and for the historical record, Darwin was shown to have been thinking about the species question, and noting down what he had thought about it, several years before Edward Blyth had written his articles in the *Annals*. Significantly, Darwin made no acknowledgment of Blyth's contributions.

Painstakingly, Gruber highlighted Darwin's unacknowledged amendments. One after another, the familiar themes were treated: the relation between food supply and population; extinction and divergence; the struggle for existence; the super-fecundity of nature; selection; variation; the law of the succession of types and biogeography. All were treated by Darwin as if they had appeared in exactly the same form in his original *Beagle* journal.

The difference Gruber found between the 1839 version and the 1845 version was simple and yet profound. He commented: 'The 1839 version gives a great deal of this evidence. The 1845 version is far more suggestive of the relation of these facts to evolutionary thought.'

Gruber also showed how much further developed was Darwin's version of his Galapagos discoveries in the 1845 edition, after using the time between 1839 and 1845 searching out new information, asking new questions and finding people who could answer them and almost completely rewriting the natural history of the Galapagos for the new version of his journal.[8]

To read the second edition without knowing these facts would inevitably have led the unsuspecting reader to credit Darwin, who was still in his twenties when he wrote the original notebooks, with a knowledge of the evolutionary process that could only have been gleaned years later. People understandably credited Darwin with phenomenal insights into the natural world and the vexed species question.

In the 1839 edition, Gruber noted, the natural history of the Galapagos had essentially passed Darwin by. When the *Beagle* was visiting the Galapagos in 1835, the Vice-Governor, Nicholas Lawson, had called Darwin's attention to the fact that the different islands of the Galapagos were inhabited by different sets of creatures. Lawson had told Darwin that the tortoises from the different islands had differing features, and that he could tell from which island any one of them had been brought.[9] This statement had no impact whatsoever on Darwin and Lawson's insights did not make it into the 1839 edition. By 1845, however, Darwin was an expert with a proprietorial interest in the species question, and his new ideas were folded back into the crucial context of the limited notes he made when he first visited the Galapagos. He completely rewrote his original Galapagos entries to take in the new ideas and information he had gleaned from Lawson and other specialists between 1837 and 1845, giving a distorted picture of how the Galapagos had struck him on the voyage ten years before.

Charles Lyell, on whose ideas many of his protégé's notions had been built, was one of the few men to whom Darwin could have confided the outlines of the theoretical ideas he was about to write into the new version of his journal.

Shortly after the publication of the 1845 edition, an eagle-eyed contemporary noted at least one difference from the 1839 version, though he did not, it seems, spot the obvious alterations to the Galapagos Islands account. In three editions of the *Gardeners' Chronicle* published between 9 August and 4 October 1845, its editor, John Lindley, referred to one of the key revisions Darwin had made in the later edition. Lindley pointed out that Darwin had added a very clear statement of the Malthusian principle of the relation between food supply and population growth to a long passage on the extinction of mammals in South America. Since Darwin only admitted to reading Malthus for the first time in October 1838, this new passage suggested that Malthus's guiding principle was known to Darwin years before he first stumbled across the philosopher.

Darwin was remarkably unfazed that Lindley had spotted his addition without attribution. In a letter to Charles Lyell shortly afterwards, Darwin wrote: 'I was much pleased by Lindley picking out my extinction paragraphs and giving them uncurtailed'.[10] This suggests two things: firstly, Darwin knew that few people were going to take the trouble to compare the two editions; and secondly, that Lyell was not likely to admonish Darwin for not giving a dated source for this new information.

The one element of Darwin's theory that Gruber did not find in either the 1839 or 1845 versions of the journal was natural selection. Why he should have resisted planting this in an appropriate place, along with all the other elements of his evolutionary thinking, Gruber had no idea:

> I do not think there is any explicit, even if veiled, expression of the unitary idea of evolution-through-natural-selection in the [1845] version of the journal. As we have seen, there is much on the theme of wholesale extermination of species, especially as a function of changes in food supply. But this is not quite the same idea as a bias acting within a species, promoting the survival of some variants and not others thus making for a change within the species... Clearly, although he expressed himself rather fully on many subjects when writing the journal, this was one area in which he held almost everything back.[11]

The truth is that Gruber, unlike Eiseley, doubted that Darwin had grasped the idea of natural selection as a positive force for evolution by 1845. He commented, 'As a conservative force in nature, the idea of natural selection had a long history before Darwin. Darwin's problem was not so much to discover the idea in the first place as to discover its significance for evolutionary theory'.

When Gruber brought out a revised edition of his book in 1981, he altered his claims about Darwin's apparent perfidy, arguing that the insertions were more likely to be the result of Darwin's originality than a deliberate falsification. However, the cat was out of the bag, and yet another academic had questioned Darwin's scholastic methods.

Gruber seems to have failed to consider that Darwin had still not grasped the fact of divergence with modification when he revised his journal notes in 1845, a year after completing his essay. Like so many others, Gruber had been convinced by the claims of Darwin's biographers that Darwin by then understood the natural forces operating to cause widespread species change. Yet Darwin had no such understanding.

It was to be a further thirteen years before he understood the full implications of the force of divergence with modification in the development of new organic forms and that enlightenment, when it struck, came neither from his work on barnacles nor from experiments on seed migration in salt water. It came instead from a totally unexpected direction.

CHAPTER 7

New kid with a net

WRITING to the *Zoologist* magazine in 1847, a correspondent commented on the fact that no specimen of the beetle species *Trichius fasciatus* had been caught for twenty years.[1] In the next edition, there was a one-sentence letter sent from an address in South Wales: 'Dear sir, I took a single specimen of this beautiful insect on a blossom of *Carduus heterophyllus* near the falls at the top of Neath Vale'.[2] With this piece of arcane information, Alfred Russel Wallace announced himself to London's scientific elite.

Wallace was born in January 1823 in Usk in South Wales. When he was five, family circumstances forced a move across the English border to the town of Hertford, about fifty miles north of London, where he attended school until he was fourteen. He then returned to Wales as a trainee assistant to his brother William, a land surveyor who re-drew parish and property boundaries.[3] It was at this time, as he worked in the open heathland of mid-Wales, that Wallace first exhibited an interest in natural history and a fascination with science, and while he was based in Kington, he attended the new Mechanics' Institute.

When he was eighteen, Wallace and his brother moved to the town of Neath near the coast of South Wales, and established themselves as independent surveyors. It proved a successful move, but Wallace's mind was absorbed by natural history rather than plans for new buildings, and he was encouraged by the members of the many scientific societies already set up in and around Neath. Wallace lectured on basic science at the Neath Mechanics' Institute, was curator of the Neath Philosophical and Literary Institution Museum, and attended meetings at the Swansea Royal Institution on zoology, chemistry and geology.[4]

In 1843, the Neath Library had 3,700 volumes representing a wide range of thinking on almost every aspect of science and exploration. Among them was Lyell's *Principles of Geology*, but Wallace later described Darwin's 1839 version of his *Beagle* journal and Humboldt's *Personal*

Narrative of Travels to the Equinoctial Regions of the New Continent as 'the two works to whose inspiration I owe my determination to visit the tropics as a collector'. The publication in 1847 of *A Voyage up the Amazon* by W. H. Edwards prompted Wallace to decide on a destination.[5]

Wallace, who had no strong philosophical beliefs, was intellectually fearless. When he was still only nineteen, his interest in zoogeography (the geographical distribution of animals around the world) was great, and his mind unprejudiced by religious preconceptions. 'To what ridiculous theories will men of science be led by attempting to reconcile science to scripture?' he asked, when belittling the ideas of one contemporary giant of natural philosophy.[6]

In 1844, he took a job in Leicester, teaching reading, composition and arithmetic to younger students, and lecturing older students in the principles of drawing and surveying. Two incidents at the town library in Leicester determined the course of the rest of his life. First, he became aware of the ideas of Thomas Malthus; second, he met Henry Walter Bates, who was to become a lifelong friend.[7]

Wallace was drawn to any place of books. His father had worked as a librarian in Hertford, where Wallace had received his basic education, and books would have been all round him in those early years. In the Leicester town library, he again found Humboldt's *Personal Narrative*; but it was Malthus's *Population* that was to have the greatest impact. Many years later, he remembered:

> I greatly admired [it] for its masterly summary of facts and logical induction to conclusions. It was the first work I had yet read treating any of the problems of philosophical biology, and its main principles remained with me as a permanent possession, and twenty years later gave me the long-sought clue to the effective agent in the evolution of organic species.[8]

With Henry Walter Bates he shared an enthusiasm for collecting, which became a lifelong obsession for them both. As an experienced contributor to learned journals (he had already published an article in the *Zoologist* on beetles frequenting damp places), Bates taught Wallace how to collect beetles, and it was their expeditions to sites around Leicester and beyond that inspired them to become professional collectors. They reasoned that the specimens they caught would allow them to work towards a solution to the mystery of how species originate.[9] Moreover, by selling surplus specimens to societies and individual collectors they could pay their expenses while abroad.

In October 1844, a book entitled *Vestiges of the Natural History of Creation* had been published by an anonymous author, who claimed: 'My sincere desire in the composition of this book was to give the true view of the history of nature, with as little disturbance as possible to existing beliefs, whether philosophical or religious'.[10] It caused consternation among natural philosophers and the clergy; the theory it proposed infuriated them. The author wrote:

> The idea which I form of the progress of organic life upon the globe ... is, that the simplest and most primitive type, under a law to which that of like-production is subordinate, gave birth to the type next above it, that this again produced the next higher and so on to the very highest, the stages of advance being in all cases very small – namely from one species only to another so that the phenomenon has always been of a simple and modest character.[11]

This emphasis on progress (revealed by the fossil forms discovered in successive geological strata) was in deliberate opposition to the concept of non-progressionism supported by Lyell, which argued that environmental change alone explained organic change since the history of the earth is cyclical and the history of the organic world is a parallel series of cycles. 'When the current cycle of an earlier geological age returns, then iguanodons, ichthyosaurs and pterodactyls, which were perfectly suited to the ancient climate, will again inhabit the earth,' Lyell argued.[12]

Vestiges was denounced by churchmen and criticised by other scientists. Its author was identified, forty years later, as Robert Chambers, one of the most successful publishers in Britain. He had chosen anonymity because of the possible effect of outraged public opinion on his business. It proved to have been a very wise move.

Wallace, eager and excited, wrote to Bates from Neath, where he had returned early in 1845. 'Have you read [the book] or is it out of your line?'[13] Bates must have shown less enthusiasm for *Vestiges* than Wallace, because in his next letter Wallace wrote:

> I have rather a more favourable opinion of *Vestiges* than you appear to have. I do not consider it a hasty generalisation, but rather as an ingenious hypothesis strongly supported by some striking facts and analogies, but which remains to be proved by more facts and the additional light which more research may throw upon the problem.[14]

However, Wallace's scientific enthusiasm was out of step with the class system of the time. Only gentlemen of leisure were thought to be of the right calibre to be natural philosophers. A working man, almost by definition, could not be classed a gentleman. Indeed, Wallace's period at Neath was marred by his apparent failure to gain the acceptance of local natural history enthusiasts.

As his collecting continued and his scientific knowledge increased, Wallace's substantial collection of beetles was borrowed by Lewis Weston Dillwyn, acknowledged to be the foremost Welsh naturalist of his day.[15] Dillwyn used Wallace's collection to impress the members of the British Association at their annual meeting, which in 1848 was held in Swansea in South Wales. Dillwyn's praise of the collection was generous, and he acknowledged Wallace's ability, but he did not welcome him into his circle of natural philosophers.[16] Wallace's status as a trainee surveyor and his nascent socialist ideas probably did little to enhance his appeal to those at the upper reaches of the local social hierarchy. It was a problem he was to encounter all his life.

In the autumn of 1847, Wallace wrote to Bates to suggest a collecting expedition. *Vestiges* was still playing on his mind, and the idea of the collecting trip occurred to Wallace after he had been studying at the British Museum, with its overwhelming displays of beetles and butterflies. He became dissatisfied with his own collection and wrote:

> Little is to be learnt by it. I should like to take some one family to study thoroughly, principally with a view to the origin of species. By that means I am strongly of the opinion that some definite results might be arrived at. Can you assist me in choosing one that it will be not difficult to obtain the greater number of the known species?'[17]

In the end, it was a Mr Edward Doubleday of the British Museum who assured the two men that the fauna of northern Brazil was little known, and that by serious collecting they should be able to cover the costs of their expedition.[18] Wallace's proposal, Bates later wrote, was 'to make for ourselves a collection of objects, dispose of the duplicates in London to pay expenses, and gather facts towards solving the problem of the origin of species'. It was a subject on which they had 'conversed and corresponded much together'.[19]

Wallace and Bates met in London in March 1848. Here, they booked their passage to Brazil, significantly diminishing the £100 Wallace had saved during his time in Neath. They also agreed that a Mr Samuel Stevens, who lived close to the British Museum, would act as their agent, receiving and selling their specimens, and handling their income.[20] Stevens knew his business well and already had collectors operating in

New Zealand, Australia, India and South Africa. It was the great age of collecting and the two novices needed someone they could trust implicitly.[21]

Over the years, Stevens not only disposed of their duplicates at the best prices, but he also took charge of Wallace's personal collection in London, dealt with insurance, sent out cash for the two men and despatched supplies that they needed in the field. He also corresponded regularly and fully on the sale of each collection, the reaction of collectors to the novelties each batch contained, how other 'fly-catchers' were progressing and matters of general scientific interest.

Usefully for Wallace, Stevens had connections with several of the scientific societies that had sprung up over the previous twenty years. He was known and respected by members of the Entomological Society, the Geological Society and the Zoological Society. He sometimes took the opportunity to read letters sent to him by his collectors at society meetings and even arranged displays of their latest consignments for the inspection of members. Samuel Stevens was to prove exactly the right agent for Wallace, who had set himself such an ambitious goal.

Bates and Wallace left Liverpool for the mouth of the Amazon on 20 April 1848. They were the only two passengers aboard the *Mischief*, a small, fast sailing vessel. Five weeks later they arrived in the province of Para, where they set out to collect insects, birds, mammals and land snails.[22] Over the next four years, working separately from Bates in order to double the number of new species collected, Wallace created what he considered to be the finest collection of Amazonian species anywhere in the world.

Eventually, on 12 July 1852, Wallace (who was suffering from a fever) supervised the loading of his precious collection aboard the brig *Helen*, which was bound for England with a cargo of India rubber, cocoa and other local products.[23] Eighty days later, Wallace stepped ashore at Deal in Kent, in the southeast of England, carrying only his watch, some drawings of fishes and a portion of his notes and journals. Less than one hundred miles off the coast of Brazil, Wallace and the rest of the ship's company had watched in horror from lifeboats as the *Helen*, 'deck and cargo one fierce mass of flame', was destroyed by fire.[24]

Wallace had lost his most recent collections from one of the wildest (and, at that time, least-known) parts of South America. The fire had consumed examples of ten species of river tortoises (many of which Wallace believed to be previously unlogged), more than one hundred species of little-known fishes, his private butterfly collection (with hand-drawn illustrations) which contained at least a hundred new and unique species, a number of curious beetles, several species of ants in all their different states, complete skeletons of an ant-eater and cowfish, a small collection of living monkeys, parrots, macaws and other birds, and one entire palm-tree leaf, fifty feet in length, which he had hoped would form a fine object in the botanical room at the British Museum.[25]

The loss must have been heartbreaking, but Wallace offered few clues to his feelings on the long journey back to England. He later wrote:

> All my private collection of insects and birds since I left Para was with me, and comprised hundreds of new and beautiful species, which would have rendered (I had fondly hoped) my cabinet, as far as regards American species, one of the finest in Europe.[26]

The value of the loss for Wallace was not monetary; nor was it to do with his reputation as a collector. He and Bates had gone to the Amazon to solve the puzzle of the origin of species by examining the relationship between patterns of affinity and distribution among closely related species. Now, the evidence for such affinities and boundaries of distribution that he might have discovered by close comparison had gone down with the *Helen*. Wallace knew he was going to have to start again.

However, not everything had been lost. During those four years in Brazil, as a result of Stevens's connections, Wallace's name had become familiar to regular readers of the *Annals and Magazine of Natural History*. The dealer knew his customers would be fascinated to read extracts from Wallace's letters, and so he sent them in for publication. They appeared, episodically, in the monthly journal's miscellany columns, providing interesting facts relating to the collections themselves and outlining the problems faced by a professional collector in the field. Soon after receiving the first letter and consignment from Wallace, six months after the latter had begun collecting, Stevens introduced Wallace and Bates to readers of the *Annals* as 'two enterprising and deserving young men' who had despatched seven thousand specimens 'in very fine condition and a vast number of novelties, besides other rare species … and a few shells and bird-skins'.[27]

One of Wallace's letters commented on the difficulty of collecting in an area where insects were so rare. They had found butterflies that were numerous in species, but not in individuals. Beetles were 'vary scarce'. He and Bates attributed the situation to the uninterrupted extent of monotonous forest, 'over which animal life is sparingly but widely scattered'. 'However', he added with his businessman's hat on, 'this makes a difference in the commercial value of the subjects'.[28]

Further up the Amazon at Santarém, Wallace found a place where butterflies were 'rather abundant', and where several species were previously unlogged. Insects, which had hardly been seen further down the river, were found in abundance. Beetles were 'as scarce as ever' in the sandy, dry earth, but he hoped to do better in the hills, which were near 'a thousand feet high and must, I should think, produce some'.[29]

Eighteen months into the trip, Stevens submitted extracts from two further letters. The first, again from Santarém, recorded the most beautiful butterfly Wallace had yet taken. 'It is very difficult to capture, settling almost invariably high up in trees ... by means of persevering with it every day for near a month I have got a good series'.[30] The second was sent four months later, in March 1850. Wallace was now a thousand miles up the Amazon at Barra de Rio Negro, one of its main tributaries.

> Insects [are] exceedingly scarce here at this season, it being almost impossible to get half a dozen in a day worth bringing home. Birds too are equally scarce, so I resolved on a short trip up the Rio Negro to where the Umbrella chatterers are found. I spent a month there, and being fortunate in finding a good hunter, have got a small but pretty good collection of birds, considering the season... Any newspapers or scientific periodicals you can send me will be particularly acceptable.[31]

Bates, who had met Wallace at Barra to decide on their next individual expeditions, wrote to Stevens in December 1850:

> [Mr Wallace] ... is now in a glorious country, and you must expect great things from him. In perseverance and real knowledge of the subject, he goes ahead of me and is worthy of all success.[32]

Wallace, however, found the country less than 'glorious'. Stevens heard from him a month after getting Bates's letter. 'I only got a dozen *gallos* [a species of bird], whereas I had expected in less time to have secured fifty. Insects there were none at all; and other good birds excessively rare.' He indicated to Stevens that he was about to travel up two further tributaries,

> not so much for my collections, which I do not expect to be very profitable there, but because I am so much interested in the country and the people that I am determined to see and know more of it, and them than any other European traveller. If I do not get profit, I hope at least to get some credit as an industrious and persevering traveller.[33]

Between Stevens's judicious placing of letters in the *Annals* and his appearances at society meetings, it would not have escaped many members of the natural history community in London that Wallace was inclined to theorise about zoology and entomology. There would have been many who thought this inappropriate for a man who made his living catching butterflies and beetles. As someone without an independent source of income, Wallace could not hope to be a member of the scientific societies he sought to address.

Wallace came back from the Amazon with his mind full of vital observations, but with the great puzzle of the species question still unsolved. The puzzle was to remain baffling for years, but Wallace's observations in Brazil lent themselves to some radical statements about the geographical distribution of animals.[34] Wallace was convinced that animals tended to exist in small, local groups, often occupying territories with distinct boundaries. He also believed that the Amazon and its tributaries formed boundaries for several animal groups, particularly monkeys, and that certain species never passed these boundaries, even if those species lived on or in the riverbank itself.

This need to establish the geographical distribution of species was familiar to Wallace because of Lyell's insistence in his *Principles* that:

> Next to determining the question whether species have real existence, the consideration of the laws which regulate their geographical distribution is a subject of primary importance to the geologist. It is only by studying these laws with attention, by observing the positions which groups of species occupy at present and inquiring how these may be varied in the course of time by migrations, by changes of physical geography and other causes, that we can hope to learn whether the duration of species be limited, or in what manner the state of the animate world is affected by the endless vicissitudes of the inanimate.[35]

Before Wallace returned to London in 1852, not much was known about why or how species were distributed between geographical locations. He had himself been completely unaware of the effectiveness of physical barriers in limiting the range of all kinds of species. His growing awareness of such barriers to species diversity, apparent in notes and remarks he made in 1852 and 1853, supported the concept of evolution that he had taken with him to the Amazon basin four years earlier.

From Lyell, Wallace understood that whole continents had sunk beneath the sea or been forced up above the water. Only when he was in Brazil did he begin to realise that the mountains of Peru, Venezuela,

Guyana and Brazil, which form barriers around the edges of the Amazon basin, had once been below sea level. Forced up over vast periods of time, they had produced torrential rivers that filled the salt-water lake trapped between them with sediment, forming a plain two thousand miles long and eight hundred miles across. This plain, which was to become the Amazon basin, was significantly younger than the mountains surrounding it.[36] If the author of *Vestiges* was right, species found in the Amazon basin had to be younger than species of close affinity found in the surrounding mountainous regions.

On 23 July 1850, the secretary of the Zoological Society read out a letter Wallace had sent Stevens on the umbrella bird, a perching bird found in the tropical forests of the Americas [37] and which was subsequently printed in the 'Miscellaneous' section of the *Annals* in November 1851. Wallace might originally have written it to interest potential customers, but once again his scientific brain took over. He wrote that the crow-like bird lived only on river islands; it had not been seen on the mainland during the course of a four-hundred-mile-journey up the Rio Negro. Wallace indicated his interest in knowing of similar species. A hunter replied to inform him that a different species of the umbrella bird lived only in the mountainous sources of the river. The possibility occurred to Wallace that the upland species might well be ancestral to the lowland species since the mountains existed before the basin was formed. From traders and the local population, Wallace learned that there was a white species of the bird, but his quest to find it proved unsuccessful.[38]

In 1851 he tried again, while trying to shake off a near-fatal attack of yellow fever. He went up the river to an area that no European had penetrated before. 'I was now in the country of the painted turtle and the white umbrella bird and I determined to make a stay of at least a fortnight to try and obtain these much desired rarities'.[39] However, after two weeks he again retired, defeated and disappointed. There was no sign of a white umbrella bird. He concluded that an uplands species closely related to the island dwelling species was a myth. Then, to rub salt into his wounds, he discovered on the eve of his return journey that one species of the bird lived in both the alluvial lowlands and the granite uplands of the Rio Negro district.[40]

After all his work, Wallace was left only with his observations of the geographical distribution of one species of butterfly in one area of the lower Amazon. The specimens suggested that the species in the forest (on relatively young alluvial plains) had originated from closely related species to be found only in the adjacent uplands. Wallace based his nascent thesis of species transmutation on this single fact. If he had found the mountain species of the umbrella bird it would have served, along with his butterfly examples, to buttress his developing ideas. The butterfly observation was interesting, but on its own it was not enough. Two examples of species modification between older and younger environments would have been so much more convincing.

Back in London three days after arriving at Deal harbour, Wallace was immediately invited to a meeting of the Entomological Society and, sometime later, to one at the Zoological Society. He became a regular presence at meetings of both societies, and made the acquaintance of many zoologists and entomologists, but he was only ever tolerated as a fly-catcher.

Wallace was invited to read a paper ('On the Habits of the Butterflies of the Amazon Valley') in two instalments to the Entomological Society at its November and December meetings. He saw this as a chance to outline his controversial ideas on geographical distribution and species which had no dependence on a supreme being.

Wallace was not the only voice railing against creationism at the time. Six months earlier in March 1852, an article in a radical left-wing London newspaper called *The Leader*, written anonymously by Herbert Spencer, tipped its hat in the direction of the *Vestiges* and attacked Creationists who demanded proof of other theories when they were unable to provide proof for their own belief.[41]

> We may safely estimate the number of species that have existed and are existing on the earth at not less than ten million. Well, which is the most rational theory about these ten million of species? Is it most likely that there have been ten millions of special creations [Lyell's theory] or is it most likely that by continual modifications, due to change of circumstances, ten millions of varieties may have been produced, as varieties are being produced still? One of the two theories must be adopted. Which is most countenanced by facts?
>
> And here we may perceive how much more defensible the new doctrine is than the old one. Even could the supporters of the development hypothesis merely show that the production of species by the process of modification is conceivable, they would be in a better position than their opponents. But they can do much more than this. They can show that the process of modification has effected and is effecting great changes in all organisms subject to modifying influence. Though, from the impossibility of getting at a sufficiency of facts, they are unable to trace the many phases through which any existing species has passed in arriving at its present form, or to identify the influences which cause the successive modifications, yet they can show that any existing species – animal or vegetable – when

placed under conditions different from its previous ones, immediately begins to undergo certain changes of structure fitting it for the new conditions [...] and thus they can show that throughout all organic nature there is at work a modifying influence of the kind they assign as the cause of these specific differences – an influence which, though slow in its action, does, in time, if the circumstances demand it, produce marked changes – an influence which, to all appearance, would produce in the millions of years, and under the great varieties of conditions which geological records imply, any amount of change.'

Buoyed by the bold statement of such ideas, Wallace outlined his new thoughts to the entomologists, along with his fragile proof of the process of modification and change. Wallace did not hold back; indeed, he added fuel to the fire ignited by Spencer's article. He enumerated several species of butterfly found in the Amazon basin.[42] Some, he said, never left the forest. Others were 'generally widely distributed' and 'exceedingly productive in closely related species and varieties of the most interesting description, often having a very limited range'. Wallace then added:

as there is every reason to believe that the banks of the lower Amazon are among the most recently formed parts of South America, we may fairly regard those insects, which are peculiar to that district, as amongst the youngest species, the latest in the long series of modifications which the forms of animal life have undergone.

This brief statement must have disturbed some of the members of the Entomological Society, many of whom were orthodox believers (and likely friends of Lyell). By advocating a process of modification of existing species over time, Wallace was directly contradicting Lyell's belief that changes in the physical world might well result in the extinction of a species, but would never allow the slow transmutation of one species into another. Disparaging Lamarck's ideas of transmutation years earlier, Lyell had argued:

It is idle to dispute [the] abstract possibility of the conversion of one species into another ... when there are known causes much more active in their nature which must always intervene and prevent the actual accomplishment of such conversions.[43]

Lyell believed those species that were not already perfectly adapted to the new geological conditions would quickly die out, and new species already perfectly adapted to that environment would be created to replace them, thus maintaining a balance between organisms in the new conditions.

The minutes recording Wallace's lectures give no indication of comments on his evolutionary stance, or of any shock or outrage. The following month, Edward Newman, the retiring President of the society, praised field collectors for furthering the cause of science. He might have had Wallace in mind when he spoke about the need of members for the work of collectors in the field:

> the monographer cannot say to the collector, 'I have no need of you'; the very admission of such a thought is a stumbling block in our own way, a bar to our own progress. I wish to be understood as applying this last observation especially and emphatically to the case of the actual collector; to the man who, in whatever station of life, devotes his time, by night and day; at all seasons in all weather; at home and abroad; to the positive capture and preservation of those specimens which serve as the objects for our observations: he is the real labourer in the field, and if we would keep the lamp of our science constantly burning, it is to him alone that we can look for fuel to feed its flame'.[44]

Within two years, the fly-catcher with ideas above his station would throw more fuel onto the theoretical flames. He would challenge, directly, the foundations of the scientific hierarchy and the religious establishment.

CHAPTER 8

A land bridge too far

CHARLES DARWIN emerged, barnacle-free, in September 1854 from his eight-year self-imposed hibernation at Down House, ready to renew his search for the answer to the species question. He was probably aware of Wallace's presence at Entomological Society and Zoological Society meetings over the previous two years. Among Darwin's friends were some of the most eminent entomologists in the country, and it is also most unlikely that he had not heard that Wallace had made extremely contentious claims about species as a result of his investigations in Brazil. Darwin had friends at the Zoological Society who were staunch supporters of the traditional theory. One of these was Thomas Bell, who had identified the reptiles Darwin had brought back with him from the *Beagle* expedition.

Darwin had certainly read two of Wallace's publications and had not been impressed by either.[1] Neither *Palm Trees of the Amazon* nor *Narrative of Travels on the Amazon* had been particularly well received when they were published. Darwin complained that the *Narrative of Travels* lacked facts, ignoring the information on the geographical distribution of animals Wallace had detailed in the work. Certainly, by the time Darwin told his friend Joseph Hooker[2] that he was about to go through his old notes on species with the intention of starting again, he would have been aware of Wallace's ideas, and his determination.

Wallace was driven to solve the mystery of the origin of species in the wild; after four years in the Amazon, he was planning a trip to the Malay Archipelago. Darwin, on the other hand, intended to rely on friends and correspondents around the world and from all kinds of scientific disciplines to supply him with information. With hundreds, even thousands of facts from every facet of natural history, he intended to uncover the remaining secrets of transmutation at his home, surrounded by his growing family and within easy travelling distance of the scientific societies of London. He was still of the opinion that at any one time there was a finite number of species, and that a new species came into existence only if an environment was drastically altered by geological forces and existing species became extinct. Then some already existing species migrating to the changed environment varied away from the norm until they became perfectly adapted to the new conditions. The world was a place of harmony and balance, and there was no possibility of species change when every existing species was perfectly adapted to its geographical environment. In such a world, there could be no variation within a species, since such a variety would no longer be perfectly adapted to the environment.

By way of analogy, Darwin had envisaged a finite number of buds on the twigs of a mature tree.[3] Each year, the buds produce leaves that eventually die and fall off, at which time the tree begins to make buds for the following year. Since the tree is mature and healthy, the same number of buds will appear the following year: no more and no fewer. So, he reasoned, it was with the total number of species existing in a perfectly balanced and harmonious world. Only when geological catastrophes killed off entire species or created new environments would new species arrive, to replace those that had become extinct or to occupy (by wind- or tide-driven migration) newly-created environments, such as volcanic islands. In the absence of such upheavals, there could be no species transmutation.

During his years studying barnacles, Darwin must have thought about the fact that his great friend Hooker had offered only faint praise for his 1844 essay. The two men spent a day together at Hooker's home in November 1846, and Darwin, impatient to get Hooker's opinion of his essay, invited Hooker back to Down House. Hooker arrived at Down for a visit of ten days, and when he left he took Darwin's essay with him. In February 1847 Hooker was still reading it, and Darwin was aware that he might be pushing Hooker too hard. He begged him not to put himself out:

> do not think of my sketch; I should never forgive myself if you look at it one minute before you have leisure and idle time: only when you recommence, oblige me by relooking over the marginal headings, so as to have the *whole* in view at once.[4]

When Hooker had finally finished, he had made several notes. Some were favourable, some were not.[5] In particular, Hooker was unconvinced by Darwin's belief that wind-and tide-borne migration was the only explanation for the presence of species on oceanic islands, such as the Galapagos.

In two letters early in March 1847, Hooker offered objections to

Darwin's idea, but communicating the gist of them to Darwin was proving difficult.[6] On some occasions Darwin was ill and could not get to Hooker's home in Kew (where his father was Director of the Royal Botanical Gardens); Hooker, on the other hand, wished to leave for India with an expedition to study the flora and geology of the eastern Himalayas. When he heard of the expedition, Darwin wrote of his disappointment: 'I shall feel quite lost without you to discuss many points with, and to point out ... difficulties and objections to my species hypotheses'.[7] In the event, Hooker was the only person to read Darwin's essay during the latter's lifetime, and his less than enthusiastic attitude almost certainly put paid to any thoughts Darwin might have had about its persuasiveness.

In November 1854, Darwin, having begun once again with species, felt he needed more facts about *exactly* what happened to animals in the wild, but – unlike Wallace – he was in no position to make such observations. If he had any thoughts about Wallace's theorising to the Zoological Society and the Entomological Society about evolution, and particularly Wallace's observations about the geographical distribution of animals, he kept them to himself. By the time Wallace reached the Malay Archipelago, Darwin had written nothing worth comment, either in public or in his private notebooks, about the ideas contained in his essay (which was, by then, ten years old).

Yet he had been offered an insight that invalidated his 1844 theory. His conviction was that there could be no variation in a world where species were perfectly adapted, but his barnacle research had revealed that there was a great deal of variation in the natural world, which came about without any catastrophic geological change. Inexplicably, Darwin did not put this aspect of his barnacle research at the centre of his species work. In fact, he told Hooker in a letter in 1850 that studying species of barnacles had not had much effect on his theories of varieties.[8] Four years later, when his barnacle research ended, Darwin continued to believe that his ten-year-old ideas were valid, and was prepared to defend them against all-comers. Startlingly, despite the clues relating to variation his research had offered, he had no insight that divergence with modification was the central driving mechanism of species change.

That November, he wrote a note to himself headed 'Divergence'. As before when he had used the word in his notebooks, he employed it as a descriptive label for ways in which organisms might be classified, rather than to describe how varying organisms in nature are modified and diverge further and further from the parent type. Four years before he began writing *On the Origin of Species*, he still did not recognise divergence as a special mechanism.[9] His theory still relied on the premise that evolutionary change was possible only in newly created geological environments to which existing species had migrated, and where they changed their form as they adapted to the new environment.

One idea in particular, by now eight years old, had raised doubts in Hooker's mind about Darwin's dependence on species migration. Early in 1846, Edward Forbes, a geologist and one of Darwin's friends, had proposed a solution to the puzzle of why certain species of plants, with strong affinities, could be found on both the African mainland and the volcanic Canary Islands, hundreds of miles out into the Atlantic.[10] Forbes's explanation was very different from Darwin's belief in the migration of species from the nearest mainland, which Darwin had taken from Lyell. Forbes suggested that a land mass above sea level supporting a multiplicity of species had once connected the present area of the Mediterranean to Europe and Africa, out into the Atlantic to Madeira and the Canary Islands, and as far north as Ireland. At some time in the distant past, geological forces had caused most of this land mass to drop beneath the sea, leaving only the Atlantic islands as remnants. Inevitably, it became known as Forbes's Atlantis theory. For some zoologists and entomologists it was an interesting idea, because it accounted for the fact that certain species of land molluscs and wingless insects were to be found on the African mainland as well as on the Canary Islands and Madeira.

This idea was damaging to Darwin's theory, and he began his new assault on the species question by setting out to prove Forbes wrong. The challenge animated him. For the next two years, Forbes's theory was an obsession that absorbed much of his time and he did not take the chance to stand back to examine the basic propositions of the theory he had been wedded to for over a decade.

Darwin appealed to Hooker.[11] His best friend was no geologist, but he was unsurpassed as a botanist. His reply must have dismayed Darwin. Hooker said that he was not convinced that the Azores and Madeira were filled with European – mainly Mediterranean – plants because of a land mass that might once have united them. However, he did say that Forbes was absolutely correct to state that there were species of plants that were found only in the west of Ireland and the province of Asturia on the coast of northwestern Spain. Hooker suggested that a land mass had once connected Ireland and Spain. He told Darwin:

> I cannot account for this by any known probable laws of migration ... I am inclined to admit any theory that will appear so botanically reasonable as that proposing the existence of land between Asturias and Ireland the apparent proof of which is drawn from the fact of the very ten plants,

> which would be likely to have availed themselves of this bridge being found at its opposite ends.[12]

Hooker's opinion was always valued by Darwin, but the latter was convinced that migration was a better explanation of this phenomenon than land bridges. In March 1855, Forbes's theory was again commended to Darwin by an entomologist, Thomas Wollaston, because it could account not only for the identical species of land molluscs found on Madeira and the African mainland, but also for species of plants and insects found in both locations, which were now separated by hundreds of miles of ocean.[13] Wollaston had earlier written: 'it is impossible to deny that, so far as the Madeiras betoken, everything would go to favour this grand and comprehensive idea.'[14] Darwin again wrote to Hooker insisting that Wollaston's insects were not confirmation of Forbes's Atlantis theory.[15]

He was so determined to prove Forbes wrong that the following month he revived an old idea and began a series of experiments in an attempt to prove that seeds could survive immersion in salt water. He was trying to prove that his theory was correct, and that species of plants could have been taken to island locations by wind and tidal currents.[16] By early May, he was complaining to William Fox, his cousin, that all his migration experiments were going wrong, and he was getting out of his depth: 'all nature is perverse and will not do as I wish it'.[17]

Two weeks later, he wrote to the *Gardeners' Chronicle* to report on his experiments. He explained that the flotation of seeds by wind and tide across oceans would account for the geographical distribution of organic forms more believably than Forbes's theory of continental extensions.[18] Towards the end of May, he was insisting to Hooker that Forbes was wrong, and a week later he wrote that Forbes's idea lacked independent evidence.[19] In return, Hooker criticised Forbes, Darwin and Lyell for offering theories that were less than acceptable.[20] He told Darwin he considered Lyell's view of creation 'neither more nor less than a superstition', and that Darwin's view was believable, 'until you work back to the vital spark – or whatever you might call it: which is a fact as inscrutable as a full blown species'. He made it clear to Darwin that as far as he was concerned, Forbes's land bridges, Darwin's migration and Lyell's centres of creation theory were all unsatisfactory explanations of the geographical distribution of plants.

Two days later Darwin told M. J. Berkeley, a fellow enthusiast, that Hooker was unconvinced that his seed experiments would help to explain geographical distribution.[21] Hooker, still unimpressed by Darwin's experiments, asked how far he would attribute certain characteristics to common descent, as Forbes had done. It was a direct question regarding Darwin's belief in the possibility of transmutation, and Darwin's reply was clear and unequivocal. (It also indicated powerfully how little evidence he had on which to base his theoretical ideas.)

> You ask how far I go in attributing organisms to a common descent: I answer I know not; the way in which I intend treating the subject, is to show *(as far as I can)* the facts and arguments for and against the common descent of species of the same genus.[22]

Darwin was suddenly very much on the defensive. He could not yet prove that species migrated to oceanic islands, and he knew that his experiments must succeed if he was to continue to insist that species evolve only when they are isolated in previously barren environments. If plants and animals did not migrate to newly formed islands like the Galapagos, no organism could evolve, since in older environments all species were already perfectly adapted and existed together in harmony and balance.

Even when some of his seeds germinated after a long immersion in salt water, his mind was not settled. Still attempting to accommodate long-held beliefs, he must have felt that his scientific reputation had suddenly become a lot more precarious. One friend, Forbes (by then deceased) had an idea that was supported by another friend, Wollaston. Now, even his best and most admirable friend, Hooker, thought Darwin's ideas less than satisfactory. In addition, Darwin had been forced to admit to Hooker that his theory could not explain how new species originated.

By August 1855, his ongoing obsession with Forbes had taken up a whole year of his life. His species ideas had not progressed since his essay of 1844. He still believed that new species arose only in newly formed and isolated environments, such as volcanic islands, and that natural selection could work only in remote territories. He could not explain extinction. If species were perfectly adapted to their environment and their environment did not change, what caused some species to become extinct, when other, similar species continued to exist in that same environment?

At that very moment, an article was being typeset that was about to cause him greater consternation than anything written by Edward Forbes.

CHAPTER 9

New law from Sarawak

AS WALLACE watched, a huge butterfly sailed high above his head. Its elongated wings were black with a horizontal band of brilliant silky yellow, and its beauty and elegance were overpowering. He thought that this butterfly in flight was one of the finest sights an entomologist could witness, but he despaired of capturing one. The specimen he saw in the vast forest inland from Malacca (on the west of the Malay peninsula) sailed along at a great height, often going for a considerable distance without moving its wings, 'in a manner quite distinct from that of any other with which I am acquainted'.[1]

He was not the first to admire the butterfly, known to entomologists as a 'bird-wing' because of its wingspan, which could measure up to 8½ inches. Nearly one hundred years earlier, Carl Linnaeus, the Swedish naturalist, had examined a specimen closely resembling that observed by Wallace, but caught around two thousand miles to the east of Malacca. Linnaeus, too, had marvelled at the insect, which in his case had black and green patterns on the upper and lower surfaces of both wings. 'Of all butterflies [it is] the most outstanding and respected ... so that I doubt whether anything more beautiful among insects has been produced by nature', Linnaeus had written.[2]

Wallace was familiar with Linnaeus's description of the family of butterflies to which the specimen belonged. Before he had left for Singapore on the first leg of the expedition that would keep him in the Malay Archipelago for more than eight years, Wallace had studied five different species of the bird-wing butterfly at the British Museum. He also knew that other travellers from Europe had taken examples back to their own countries.

It must have provoked incredible feelings in Wallace to see the insect in flight rather than pinned to a board at the British Museum, but those five species, along with a book on butterflies written and illustrated by the French naturalist J. A. Boisduval,[3] allowed him an insight into almost every species of bird-wing butterfly, *Ornithoptera*, that had ever been known.

Towards the end of 1854, Wallace, who by then had travelled on to Sarawak on the island of Borneo, was surprised to be presented with another, dead specimen of the bird-wing butterfly. He saw immediately that it was not only unlike the specimen he had seen near Malacca, but it was also unlike any of the nine species listed in Boisduval's book, or the five species he had studied at the British Museum.[4] The butterfly he had seen near Malacca had black wings with yellow markings; he had seen a similar specimen at the British Museum. The dead specimen he was given at Sarawak, five hundred miles east of Malacca, had silky black wings with green markings, but it was also different from anything any other entomologist had recorded.

Various specimens of the butterfly could be found along an arc stretching from India in the west, down through the Malay Archipelago and New Guinea, to Australia in the southeast. Wallace knew immediately that this specimen was a new species. He called it *Ornithoptera brookiana*, and described it minutely. He began to draw deductions about its importance in terms of geographical distribution:

> This magnificent insect is a most interesting addition to the genus *Ornithoptera*. The green-marked species have hitherto been found only in N. Australia, New Guinea and the Moluccas, and all those yet known so much resemble each other in their style of marking, that most of them have been considered as varieties of the original [which Linnaeus had found in 1758]. Our new species is therefore remarkable on two counts: first, as offering a quite new style of colouring in the genus to which it belongs; and secondly, by extending the range of the green-marked *Ornithopterae* to the N.W. extremity of Borneo. As it has not been met with by the Dutch naturalists, who have explored much of the S. and S.W. of the island, it is probably confined to the N.W. coast. My specimen (kindly given me by Captain Brooke) came from the Rejang river; but I have myself once seen it on the wing near Sarawak. I have named it after Sir J. Brooke, whose benevolent government of the country in which it was discovered every true Englishman must admire.[5]

What must have puzzled Wallace greatly, given that he had been in the Malay Archipelago for less than eight months, was the unexpected distance between the habitat of the Borneo bird-wing butterfly and New Guinea, two thousand miles away, where a species of close affinity had been taken many years before.

Wallace knew that Europeans were familiar with the central islands of the archipelago, including Borneo; they had explored them for hundreds of years, and yet the green bird-wing butterfly had never before been sighted there. Previously, all known examples of the green bird-wing had been found in the islands west and north of New Guinea, where the Dutch, and the Portuguese before them, had established the spice trade and sent back specimens of the green bird-wing for collectors in Europe.

It must have struck Wallace that there was a huge area of the Malay Archipelago between the two different species of the bird-wing butterfly where no species of *Ornithoptera* existed.[6] If the range of the bird-wing butterfly had been established, on his own evidence, as being between Borneo and New Guinea, why was the whole central area – hundreds of thousands of square miles – dotted with islands seemingly empty of bird-wing butterflies?

Wallace would have realised that he was faced with a geographical distribution gap between different species that had emerged from a common ancestor. Somewhere between the two closely allied bird-wing forms, an ancestral species had once existed from which the two species he was comparing had sprung. Somehow, they had been modified: changed in appearance and developed in different ways. It was impossible to identify a common ancestor; the species from which they had both derived had become extinct. Wallace was certain that only this explanation could account for the existence of such isolated but closely related groups of species.

Wallace's new idea – that a species could have arisen only from a contiguously pre-existing species – clashed directly with Lyell's ideas. Lyell would have said that Wallace's *Ornithoptera brookiana* would have been created separately for the environment where it had been discovered, without reference to any other existing species of bird-wing.

If different species of the bird-wing butterfly could be found in areas more than a thousand miles apart, what was there to prevent other closely related species of animals existing in geographical areas thousands of miles apart? Wallace began to see that there were striking common characteristics in species of birds that could be found in the Malay Archipelago and also in the Amazon basin, more than ten thousand miles away. In the Malay Archipelago, he found brown-backed trogons (small birds found in tropical forests) which had a closely related green-backed cousin in the Amazon. Among the families of butterflies he was able to connect, the *Euploeas* of the east and the *Heliconidae* of South America were, it seemed to him, members of the same extended family.

It all seemed to fall into place. All these exemplary species at some time in the distant past must have had common ancestors. It was a far simpler idea than Lyell's theory, which posited that each new species had been created by some metaphysical force to occupy a place on earth to which it was already perfectly adapted.

Less than a year after leaving England, Wallace began to write out his ideas. Some years later, he described the paper he completed in Sarawak as 'my first contribution to the great question of the origin of species'. He explained:

> Having always been interested in the geographical distribution of animals and plants ... and having now myself a vivid impression of the fundamental differences between the Eastern and Western tropics ... it occurred to me that these facts had never been properly utilized as indications of the way in which species had come into existence. The great work of Lyell [on fossils] had furnished me with the main features of the succession of species in time, and by combining the two I thought that some valuable conclusions might be reached.[7]

In a small house at the mouth of Borneo's Sarawak river during the wet season, alone with one Malay boy as cook, Wallace had time to look over his books and ponder the implications of his new ideas. Within a short space of time, he laid out for the natural philosophers back in Europe a treatise on species that was light-years ahead of anything that had been published up to that time.[8]

He used Lyell's *Principles*, and allied them to his own observations and close examination by dissection of the structure of specimens. He was convinced that, along with the constant changes in the earth's crust, there must also have been a gradual alteration of the forms of organic life. The earth's crust had undergone so many changes over time that all species that once existed had become extinct, only for their places to have been taken by new species. It followed, therefore, that such a process would have happened between every subsequent geological time period since life began.

Having tested these propositions against all the 'newly ascertained facts with which I have become acquainted, or have been able to observe myself', Wallace felt he had enough evidence for a new law. Considering the extreme complexity of the subject he was dealing with, Wallace's Sarawak Law, as it has come to be known, was absolutely precise. It stated simply: 'Every species has come into existence coincident both in space and time with a pre-existing closely allied species'.

It was a revolutionary idea. Wallace believed his law could explain the distribution of species of animals and plants across the continents of the world, and why their emergence over time in these places depended not on God's will, but on the previous existence of similar kinds of organisms in those places. He argued:

> If the law enunciated be true, it follows that the natural series of affinities will also represent the order in which the several species came into existence, each one having had for its immediate antitype a closely allied species existing at the time of its origin.

Wallace was taking on Lyell very directly. There was no room here for Lyell's idea that when geological forces caused the extinction of a species, some First Cause created new species that were already perfectly adapted. Wallace felt that Lyell was absolutely wrong to depend on such religious concepts. Perfect adaptation was not an answer. Wallace was very clear about how species modified and eventually diverged:

> It is evidently possible that two or three distinct species may have had a common antitype, and that each of these may again have become the antitypes from which other closely allied species were created. The effect of this would be that so long as each species has had but one new species formed on its model, the line of affinities will be simple, and may be represented by placing the several species in direct succession in a straight line. But if two or more species have been independently formed on the plan of a common antitype, then the series of affinities will be compound and can only be represented by a forked or many branched line.
>
> Now, all attempts at a Natural classification and arrangement of organic beings show that both these plans have obtained in creation. Sometime the series of affinities can be well represented for a space by a direct progression from species to species or from group to group, but it is generally found impossible so to continue. There constantly occur two or more modifications of an organ or modifications of two distinct organs, leading us on to two distinct series of species, which at length differ so much from each other as to form distinct genera or families. These are the parallel series or representative groups of naturalists, and they often occur in different countries, or are found fossil in different formations. They are said to have an analogy to each other when they are so far removed from their common antitype as to differ in many important points of structure, while they still preserve a family resemblance. We thus see how difficult it is to determine in every case whether a given relation is an analogy or an affinity, for it is evident that as we go back along the parallel or divergent

> series, towards the common antitype, the analogy which existed between the two groups becomes an affinity.

Quite sure of his argument, Wallace summed up:

> The process of extinction within lineages from a common ancestor together with the gradual modification of species is both necessary and sufficient to explain all of the affinity relationships evident in the organic world.

A little later in his paper, Wallace suggested the complexity of attempting to trace relationships in order back to a common antitype in nature would be as intricate as attempting to work out the complicated branching of the twigs of a vast oak, or the vascular system of the human body. He added:

> if we consider that we have only fragments of this vast system, the stem and main branches being represented by extinct species of which we have no knowledge, while a vast mass of limbs and boughs and minute twigs and scattered leaves is what we have to place in order, and determine the true position each originally occupied with regard to the others, the whole difficulty of the true Natural System of classification becomes apparent to us.

Wallace declared that while his theory might appear to some readers essentially a theory of progression:

> it is in reality only one of gradual change [but one which] claims a superiority over previous hypotheses on the ground that it not merely explains, but necessitates what exists. Granted the law, and many of the most important facts in Nature could not have been otherwise, but are almost as necessary deductions from it, as are the elliptic orbits of the planets from the law of gravitation.[9]

Wallace's understanding of divergence as a driving mechanism for change runs through the entire article. However, this was not the first time he had indicated the existence of some such law or principle in nature.

In 1854, the miscellany section of the *Annals* had reprinted an extract from *Narrative of Travels on the Amazon and Rio Negro*, the book Wallace had written in 1853 following his experiences in South America. Under the title 'Habits of Birds', Wallace had written:

> In all works on Natural History, we constantly find details of the marvellous adaptation of animals to their food, their habits, and the localities in

> which they are found. But naturalists are now beginning to look beyond this, and to see that there must be some other principle regulating the infinitely varied forms of animal life. It must strike everyone, that the numbers of birds and insects of different groups, having scarcely any resemblance to each other, which yet feed on the same food and inhabit the same localities, cannot have been so differently constructed and adorned for that purpose alone.[10]

Now, less than two years later, Wallace must have been confident that the law he had newly formulated was the announcement of the discovery of that principle.

Wallace sent his paper to the editor of the *Annals and Magazine of Natural History* in London.[11] It was published on the first day of September the same year.[12] The reaction was a deafening silence, apart from a comment to Samuel Stevens indicating that Wallace should curtail his theorising and concentrate more on his collecting.

Perhaps it was the way Wallace had expressed his certainty. Perhaps it all seemed too strong in the columns of a magazine where gentlemen naturalists – rather than fly-catchers – were expected to air new ideas. Perhaps it was because Wallace had ignored the belief that the natural sciences were not for amateurs making their living by collecting in forests. Perhaps it was because the ideas in the article were simply too advanced for the readers of the *Annals*. Perhaps it was simply a case of snobbery.

Whatever the reason, when the most profound, logical and well-argued essay on evolution yet published dropped into the letterboxes of London's natural philosophers in September 1855, it was totally ignored.[13] There was no outrage, no discussion, no comment and no reaction. It was as if the Sarawak Law had never been written.

Yet Wallace had advanced thinking on the origin of species effortlessly and with a profound clarity of expression. Directly and without fear, he had dismissed in a few paragraphs the metaphysical ideas of Lyell and argued that logic should determine the limits of the debate.

Whatever the ultimate effect on Charles Lyell, the Sarawak Law caused a huge problem for Darwin. It was completely opposed to his conviction that because of the perfect adaptation of species to their surroundings, species change could happen only in isolated, newly formed environments where, by definition, there could be no fossil remains or pre-existing ancestral types.

Once the Sarawak law had been published, it was not long before Darwin's thoughts about the species question, as recorded in his notebooks, began to move in an entirely different direction.

CHAPTER 10

Silence gentlemen, please

WHEN MEMBERS of the various scientific societies in London read Wallace's Sarawak paper, they must have been amazed to discover that not only was Wallace advocating an entirely new causality between extinct and living species but that he was also using Lyell's geological discoveries to highlight deficiencies in the great man's thinking.

Thomas Wollaston, the entomologist, remained silent with regard to Wallace's thoughts about the geographical distribution of butterflies. Hooker – who was Assistant Director, under his father, at the rapidly expanding Royal Botanic Gardens in Kew – had nothing to say about the implications of Wallace's ideas for botany. Members of the Zoological Society were apparently unaffected by Wallace's claims, which might be thought to apply to families and genera of large mammals, specimens of which were to be found on continents thousands of miles apart.

Wallace's first great contribution towards solving the puzzle of the species question seemed to pass over the heads of everyone. But that was not how Charles Darwin saw it. Having strived to convince Hooker that his theory was superior to Forbes's, he now saw his migration theory coming under renewed attack. Hooker would have been able to see that his theory was vulnerable to the Sarawak Law. It asked many questions of Darwin's thinking on species, and particularly on the question of geographical distribution.[1]

To discover that Wallace was so much in control of the arguments, with knowledge based on years of observation in the tropics, must have caused Darwin some distress. Suddenly he was being presented with evidence from the Amazon and the Malay Archipelago that the geographical distribution of organic forms around the world might be explained by species variation and extinction and not, as he had so long believed, by migration to newly-formed geographical regions.

Yet nothing in Darwin's correspondence, collated and published by his son Francis after his death, directly mentions the existence of Wallace's

Sarawak paper. There was no reference to it in his daily journal, which was found sixty years after his death, or in his short autobiography. Despite the fact that he had been aware of Wallace's longstanding interest in discovering how species originate since his return from the Amazon, Darwin saw no reason to pass comment on Wallace's recent ideas.

When the Sarawak Law article appeared in the *Annals*, he had been still fixated on his attempts to prove his seed flotation and germination theory. By the end of November 1855, only three months after the article was published, Darwin had downgraded his seed flotation experiments and begun his new study of the cross-breeding of domestic animals to understand how an advantageous degree of variation might be achieved and exploited to cause modification in species. With the thought that it might throw light on how nature brings about similar changes in creatures in the wild, Darwin had decided to study man's breeding of pigeons.

But in the middle of April 1856, Darwin was visited at his home by Sir Charles Lyell.[2] Darwin's recollection of that visit as an old man was simply that Lyell urged him to write down his ideas on evolution and get them published. Darwin recorded that he had taken Lyell's advice and immediately began writing what he always referred to afterwards as his 'big species book', leaving history to understand that Lyell's visit was pure serendipity, and his warning simply a mentor's advice to a protégé, or one friend's concern for another's predicament.

In 1961, however, a new light was shone on the vacillating state of mind of Charles Darwin in that early spring of 1856. A notebook came to light in a most unexpected place, unearthed by a Darwin enthusiast who must have felt dismayed by his own discovery. It seemed to contradict the authorised version of the story that had been shaped for posterity by Darwin's son Francis.

CHAPTER 11

Unwelcome discovery

LEONARD WILSON, a Professor at Yale, was collecting material for a book he had long wished to write about the life and career of Sir Charles Lyell. Working through documents kept at Kinnordy House, the Lyell family home in Scotland, he came across a set of manuscript notebooks missed by previous researchers.[1] They recorded Charles Lyell's interest in the species question. On the very first page of the first book was a single word: 'Wallace'.

The first entries in the book were dated 28 November 1855, nearly three months after the publication of Wallace's Sarawak Law, which Lyell had missed because he had been visiting South Africa. He had been back in London only two days when he opened that notebook. In between, he had studied Wallace's article, and it seems to have thrown into confusion much of Lyell's thinking. He was obviously shocked by Wallace's suggestion that all species, including by implication man himself, were the offspring of pre-existing, closely allied species. Lyell confided in his notebook that he believed the puzzle about the origin of species was in the process of being unravelled by Wallace.[2]

When, five months later in April 1856, he met Darwin in Kent, he wanted to discuss the possible impact of Wallace's Sarawak Law on Darwin's own ideas, which Lyell had never read for himself.[3] Because of his concern for his protégé's scientific priority, Lyell insisted, after listening to Darwin, that he publish his ideas on species without delay or risk losing out to Wallace.

Lyell might also have had his own agenda. Wallace used Lyell's own *Principles of Geology* to support the Sarawak Law, dismissing Lyell's belief that new species were the work of a Creator. Wallace had first outlined these arguments two years previously, in his two lectures to the Entomological Society. Wallace's confidence in his own ideas must have given Lyell some cause for concern.

In April 1856, when he returned home from his visit to Darwin, he made further entries in his notebook based on what Darwin had told him about the latter's work. On April 16, he wrote 'With Darwin: On the Formation of Species by Natural Selection – (Origin Query?)' Darwin had obviously been convincing, because Lyell noted: 'The reason why Mr. Wallace introduction of species most allied to those immediately preceding in Time ... seems explained by [Darwin's] Natural Selection Theory'.[4]

Darwin, it appears, had been able to assure Lyell that Wallace's ideas held no concern for him; that they were only a pale shadow of his own ideas, which he had written out at great length in 1844. Lyell, seemingly, was not given the impression that Darwin had spent much time analysing Wallace's article. Lyell jotted down the younger man's thoughts, and accepted at face value Darwin's personal assurances that his ideas of 1844, and particularly his concept of natural selection, were a long way ahead of the ideas of Wallace.

In a letter following his visit, Lyell was insistent that Darwin should outline his theory in public, despite the fact that he, too, now had reservations about Darwin's idea of species migration across oceans to new environments.[5] He informed Darwin of a lecture he had attended, at which a Swiss palaeontologist had made out a good case for a land bridge between Madeira and the mainland of Africa. When they met, Lyell had been convinced that Darwin really did have a coherent theory, even though migration was at its heart. Lyell pleaded for Darwin to reach out and grasp the prize of academic priority: 'I wish you would publish some small fragment of your data, *pigeons* if you please, and so out with the theory and let it take date and be cited and understood'. Darwin replied:

> With respect to your suggestion of a sketch of my view, I hardly know what to think, but will reflect on it; but it goes against my prejudices. To give a fair sketch would be absolutely impossible, for every proposition requires such an array of facts.

Darwin was effectively admitting that his ideas lacked any supporting evidence:

> If I were to do anything it could only refer to the main agency of change, selection, - & perhaps point out a very few of the leading features which countenance such a view, & some few of the main difficulties. But I do not know what to think: I rather hate the idea of writing for priority, yet I certainly should be vexed if any one were to publish my doctrines before me. [...] If I did publish a short sketch, where on earth should I publish it?[6]

It must have been a difficult time for Darwin. Stranded between the ideas of his twelve-year-old essay and his failing experiments to get seeds to germinate after being soaked for weeks in salt water, Darwin turned to Hooker.

> I very much want advice and *truthful* consolation if you can give it. I had a good talk with Lyell about my species work and he urges me strongly to publish something [...] If I publish anything it must be a *very thin* and little volume, giving a sketch of my views and difficulties; but it is really dreadfully unphilosophical to give a resume, without exact references, of an unpublished work[7].

Darwin, however, had an idea of how to get round that problem:

> In Preface I would state that the work could not be considered strictly scientific but a mere sketch or outline of future work in which full references etc should be given. Eheu, eheu, I believe I should sneer at anyone else doing this, & my only comfort is, that I *truly* never dreamed of it, till Lyell suggested it, & seems deliberately to think it adviseable. I am in a peck of troubles and do pray forgive me for troubling you.

There is no indication in any of his correspondence that Darwin had any new theory or convincing evidence in the late spring of 1856. Darwin knew that his ideas might be considered threadbare, implausible or out of date by comparison with Wallace's paper. Hooker wrote by return, in a letter since lost, agreeing with Lyell's idea that he publish something.[8] Hooker suggested that Darwin write a preliminary essay to get round the problem, but Darwin still had a problem. On May 11 he told Hooker again that it would be wrong to publish without full details. 'I begin *most heartily* to wish that Lyell had never put this idea of an Essay into my head.'[9]

On the same day, Darwin wrote a note to himself, which he headed 'Classification'. It read: 'New species must be created to some pre-existing idea or plan. But ... how according to descent can be explained this curious arrangement of all living and extinct beings'.[10] It was less than four weeks since Lyell's visit. This sudden interest in the 'curious arrangement of all living and extinct beings' had not appeared from nowhere: was it a mere coincidence that this idea was at the heart of Wallace's Sarawak Law?

It must also have been clear to him, as he prepared to write his preliminary essay, that the transmutation of migratory species in new environments formed by geological forces was not enough to explain the origin of new species. It was time for new ideas. There was only one problem: at this stage, Darwin had none.

In his journal for 14 May 1856, Darwin wrote, 'Began by Lyell's advice writing species sketch'.[11] After this date, as Dr Robert Stauffer noticed when he first trawled through Darwin's papers at Cambridge in 1959, Darwin's letters began to dwell more on problems concerning the geographical distribution of plants and animals. Stauffer observed that this was a new departure for Darwin: it was not an area to which he had paid much attention before.[12]

Patterns of geographical distribution had long struck Charles Lyell as one way of moving towards a solution of the species problem. The idea's importance had also driven Wallace in his search for valid examples during his time in the Amazon basin, and was echoed in his lecture to the Entomological Society less than three years before. Along with the idea of divergence, the geographical distribution of animals was at the heart of the Sarawak Law and could not have been missed had Darwin understood the paper thoroughly. Yet, despite his sudden interest in geographical distribution, there is no indication that Darwin admitted to any of his friends that he had paid any attention whatsoever to Wallace.

Meanwhile, in June, writing to his cousin William Fox, Darwin was less than wholly honest about why Lyell had come to see him:

> Sir C. Lyell was staying here lately and I told him somewhat of my views on species and he was sufficiently struck to suggest (and has since written so strongly to urge me) to me to publish a sort of Preliminary Essay.[13]

Notwithstanding his preoccupation with priority, Darwin still could not get Forbes, land bridges and island populations out of his head.

CHAPTER 12

Warning shots

H. LEWIS McKINNEY, a graduate student at Yale in the early 1960s, had been advised that for his doctorate thesis he might be interested in examining the work of Alfred Russel Wallace relating to the geographical distribution of animals. After initial research, McKinney was working on the premise that Wallace's contribution to the early history of evolutionary biology was far greater than had previously been realised. McKinney felt aggrieved that Wallace had been portrayed as having drawn much of his inspiration from Darwin's ideas, rather than as someone who had been a brilliant and original thinker in his own right. He believed that previous authors who had written about Wallace had diminished the originality of his thinking because they had paid too much attention to his published works, rather than visiting his original notebooks, which McKinney unearthed at the library of the Linnean Society in London.

Soon after beginning his research, McKinney learned of Leonard Wilson's discovery of Lyell's notebook with Wallace's name written on the first page. He approached Wilson, a colleague at Yale, who granted him access to the new material, even though Wilson's book on Lyell was a long way from completion.

After reading Lyell's account of events between November 1855 and the spring of 1856, McKinney realised that if he could find any evidence that Darwin, writing after September 1855, had used ideas first expressed by Wallace and not the other way round, then his doctorate thesis would be revolutionary. Within a very short time, he began to find hard evidence not only that Wallace's intellectual path to evolution was much less convoluted than Darwin's, but that most of that time Wallace's ideas had been ahead of Darwin's, rather than the other way around.[1]

McKinney knew that Wallace had harboured a determined ambition to solve the puzzle of the origin of species from an early age. He also discovered that in an orthodox world, Wallace had been an unorthodox man with an unblinkered way of surveying the natural world and seeing connections that had escaped other men.

At the Linnean Society, McKinney gave Wallace's original notebooks specific titles to allow other researchers to follow more easily the paths he was beginning to explore. His excitement mounted as he began to read how Wallace's ideas had formed and changed during the early years.

McKinney had long known of Lyell's profound influence on Darwin, and that the older man's religious beliefs had formed 'the essential fabric' of Darwin's own ideas on species and varieties. Wallace had taken a different path. He had used Lyell's groundbreaking geological discoveries to back up his own original ideas about evolution, while rejecting totally Lyell's belief about how species had been created. McKinney found that apart from Lyell, Wallace had borrowed from Chambers's *Vestiges* and, among others, the French theorist Lamarck, and that by putting all these influences together alongside his own observations, he had developed 'a unique, cogent argument' on the species question.[2]

By February 1855, McKinney realised, Wallace had:

> pinpointed the necessary problems to explore and had clearly recognized the intense struggle for existence, the ecological balance of species and the important fact that gradual change occurring over a long period may ultimately result in the origin of distinctly new species.[3]

All these ideas fused together to become his Sarawak Law, but McKinney was aware of one vital missing ingredient that had prevented Wallace from presenting to the world the fully formed theory of evolution four years before the publication of Darwin's *On the Origin of Species*. That ingredient was the application of the idea of the survival of the fittest. Without it, the Sarawak Law was a brilliant addition to the process of change in the natural world, but incomplete.

These insights were interesting, but McKinney unearthed his first major find in Darwin's formidable collection of loose papers at Cambridge. There, he found a copy of the *Annals and Magazine of Natural History* for September 1855, in which Wallace's Sarawak Law had been published. McKinney knew that Darwin had not made any reference to Wallace's ideas at the time, but the pages McKinney examined bore heavily scored lines in the margins, with notes, annotations and a simple, diagrammatic sketch. He realised that what he was reading was Darwin's own copy of Wallace's article.[4] Against almost every paragraph, Darwin had drawn straight lines in the margin, 35 in all, marking out what he obviously considered to be points of interest, at the very least.[5]

In some places there were double parallel lines, and in three places the lines were heavily scored, suggesting that Darwin had been particularly struck by these passages. Where Wallace had written 'Species of one genus, or genera of one family occurring in the same geological time [period] are more closely allied than those separated in time', Darwin, in his almost illegible scrawl, had written 'Can this be true?'[6] McKinney noted, 'Wallace's point was clearly novel and interesting to him'.

In the same archive he found a copy of the December 1855 issue of the *Annals*, to which was pinned a sheet of notes about the Sarawak Law, also in Darwin's handwriting. In that moment of discovery, the unlikely possibility that Darwin had not studied at length Wallace's article before Lyell had brought it to his attention in early 1856 was shattered.

McKinney had another reason for thinking that his discovery would prove important and highly controversial for Darwin scholars. In one paragraph, Wallace had offered his thoughts about the single-path development of species alongside forms diverging from a common antitype. In the margin against this paragraph, Darwin had drawn with dotted lines two small, simple diagrams: one representing a continuous path of development, and one diverging.[7]

McKinney commented, 'Then there is the matter of the branching lines of affinity which has obvious meaning.' It must have been an incredible moment because McKinney knew that four years after reading that article, in the first edition of *On the Origin of Species*, those two simple diagrams had been turned upside-down, redrawn, developed and slightly adapted, and became the only illustrations used by Darwin to indicate various likely paths of divergence.[8]

McKinney, by then teaching at Yale, was awarded his doctorate in 1966. A book based on his thesis was published in 1972, one year after further unexpected evidence of Darwin's awareness of Wallace's article had come to light.

Edward Blyth, the chemist from Tooting, had written to Darwin in December 1855 to ask him what he had thought of Wallace's Sarawak Law.[9] (Blyth had left England for health reasons shortly after Darwin published his first account of his years on the *Beagle*. Since 1840, he had been Curator of the Asiatic Society of Bengal in Calcutta, but had kept up an episodic correspondence with Darwin, who by then was using him freely for his expert knowledge of classification, but had still never credited him for his original ideas on how selection works on species.) Blyth was full of praise, and devoted four of the eight pages of his letter to Wallace's ideas. He admitted that he was struck by its evolutionary implications, and enlightened Darwin about various additional examples of geographical distribution with which Wallace might, if he wished, support his contentions and back up his Law.[10] His enthusiasm for Wallace's paper ran through his discussion: 'Wallace has, I think, put the matter well,' he said at one point; later, he commented: '... a trump for friend Wallace to have hit upon'. Blyth inquired about the effect of Wallace's law on Darwin. He wrote:

> What think you of Wallace's paper in the Annals? Good! Upon the whole! [...] Has it at all unsettled your ideas regarding the persistence of species – not perhaps so much from novelty of argument as by the lucid collation of facts and phenomena?

Blyth's letter was dated 8 December 1855. Darwin would have received it late in January the following year, three months before Lyell's visit. It might well have been this letter that provoked Darwin to look again at Wallace's article, since the page of notes found by McKinney was attached to the June–December volume of the *Annals*,[11] which would in all likelihood have been still around in Darwin's study when Blyth's letter arrived at Down House late in January.

In the weeks that followed Lyell's visit, Darwin must have been concerned as he realised how the views of his friends and contemporaries were changing. First it had been Forbes, then Hooker, then Blyth and now Lyell. His questioning of Darwin, and his insistence on action to get his ideas published, must have proved the last straw. Darwin could no longer escape the conclusion that if everyone he trusted was no longer persuaded by his migration ideas, and even Lyell was impressed by the ideas of Wallace, then he was going to have to find a different way forward, or he was going to be left far behind.

CHAPTER 13

Darwin without a compass

THE GEOGRAPHICAL distribution of organic forms quickly became a major concern for Darwin as he acted on Lyell's advice and began writing the short sketch of his theory, which he again called 'Natural Selection'. An air of weary desperation seeps out of his correspondence as he attempts to find ways to disprove Forbes's Atlantis theory.[1]

In his frustration, he wrote to Lyell on 16 June 1856:

> [M]y blood gets hot with passion and runs cold alternately at the geological strides which many of your disciples are taking. Here, poor Forbes made a continent to N. America... Hooker makes one from New Zealand to S. America and round the world to Kerguelen Land [an island in the Southern Ocean]. Here is Wollaston speaking of Madeira and P. Santo 'as the sure and certain witnesses' of a former continent. Here is Woodward writes to me if you grant a continent over 200 or 300 miles of ocean depths (as if that was nothing) why not extend a continent to every island in the Pacific and Atlantic oceans. And all this within the existence of recent species![2]

Lyell wrote back immediately, but had no sympathy:

> I wonder you did not also mention D. Sharpe's paper just published by which the Alps were submerged as far as 9000 feet of their present elevation above the sea in the Glacial Period and then since uplifted again [...] The littoral shells according to Macandrew imply that Madeira and the Canaries were once joined to the main land of Europe or Africa but that those isles were disjoined so long ago that most of the species came in since. In short the marine shells tell the same story as the land shells[3].

Darwin replied eight days later:

> If on account of European plants and littoral sea-shells it is thought necessary to join Madeira to mainland [then] Hooker is right to join New Holland [Australia] to New Zealand and Auckland Island (and Raoul Island to the N E) and these to S. America and the Falklands and these to Tristan d'Acunha and these to Kerguelen Land; - thus making, either strictly at the same time, or at different periods, but all within the life of recent beings, an almost circumpolar belt of land. So again Galapagos and Juan Fernandez must be joined to America; and if we trust to littoral sea-shells, the Galapagos must have been joined to Pacific Islands [2,400 miles distant] as well as to America ... In short we must suppose, probably, half the present ocean was land within the period of living organisms [...] For these several reasons and especially considering it certain (in which you will agree) that we are extremely ignorant of means of dispersal, I cannot avoid thinking that Forbes's Atlantis was an ill-service to Science, as checking a close study of means of dissemination.[4]

Forbes was an ongoing irritation. Darwin continued to argue because if Forbes was correct, the implication for his theory was disastrous. In a letter to Lyell on 5 July, Darwin wrote that if Forbes's theory was found to apply to just one island and continental mainland, then it must apply to all. He admitted that he still could not accept the idea, but he told Lyell that the essay on species he was writing would be made 'as complete as my present materials allow'.[5]

Darwin also asks Lyell for his judgment as to whether Forbes's idea might really be a possibility in the case of a now submerged continental extension between the west coast of Africa and the island of Madeira, 'because if I could honestly admit these great extensions they would do me good service'. In a letter to Hooker the same day, Darwin says he cannot accept the idea of continental extensions and he has written to tell Lyell as much. It is obvious that Darwin is unsure which way to go. He tells Hooker that the more he reflects on Hooker's botanical evidence, the more he is astounded:

> You give all the facts so clearly and fully that it is impossible to help speculating on the subject; but it drives me to despair, for I cannot gulp down your continent [which Hooker has suggested once existed in the southern circumpolar region, and which joined various islands]; and not being able to do so gives in my eyes the multiple creationists an awful triumph.[6]

In a follow-up letter he then asks Hooker if he has any examples that would support the theory that similar species in different geographical areas of the world were the result of multiple creations. He tells Hooker that even one proved case of multiple creations would smash his migration theory.[7] At this point, he was still persisting with his experiments to get seeds and plants to float, but his appetite for the experiments seemed to be diminishing.

Five days later, Darwin wrote to Samuel Pickforth Woodward (a palaeontologist who believed that many of Darwin's theories were based on inadequate evidence) to confess that he was 'growing as bad as the worst about species' and hardly had a vestige of belief in the permanence of species left in him.[8]

But he maintains his objection to continental extensions the following day when he tells Hooker, 'generally I would observe that I would admit a continental extension in any *few* cases, when the facts required it more than in the generality of cases, but it seems to me that you will have to admit continental extensions to every island whatever, & that I cannot swallow. Indeed even one *continental* extension is an awful gulp to me.' [9]

At this point, Hooker finally decided to act as a scientist rather than a friend. As July 1856 turned to August, he told Darwin that he preferred Forbes's theory of continental extensions to Darwin's migration theory. He told Darwin, 'I am quite ready to admit the gigantic difficulties in the way of Continental Extension, and I also admit that it does not explain all the facts and is no more than an idea perhaps; but it does not fly in the face of known facts in the history of distribution and the geological arguments against it are of no *proved* value. The Continental Extension may be a retrograde step, but it is no harm done, whereas the migration strikes at the root of logical induction from known facts in distribution'.[10]

This was too much for Darwin. By return, he told Hooker he could not agree with him:

> You cannot imagine how earnestly I wish I could swallow continental extensions, but I cannot; the more I think (and I cannot get the subject out of my head) the more difficult I find it. If there were only some half-dozen cases, I should not feel the least difficulty, but the generality of the fact of all islands (except one or two) having a considerable part of their production in common with one or more mainlands utterly staggers me.[11]

The blow for Darwin cannot be underestimated. After nearly twenty years of work, his best friend and adviser dismissed his crucial precondition for species transmutation in a few lines of a personal letter. Darwin's theoretical assumptions lay in tatters. If his migration theory was wrong, then the effect of natural selection working only on migratory species in newly created and isolated environments would also be wrong. In addition, since he assumed that species never changed in areas of the world where perfect adaptation applied, then he was suddenly left with no general explanation for species change.

In July, as Darwin disputed the merits of Forbes's theory in an increasingly urgent fashion, Alfred Russel Wallace had once again appeared in print in the *Annals* with new observations on species, variation and divergence. In a short article about the habits of the orang-utan of Borneo, Wallace speculated about why many animals have vestigial organs and appendages that serve no material or physical purpose. He added some thoughts about the orang-utan and the origins of man that would not have made easy reading for many natural philosophers, who still put their faith in the work of a Creator.[12]

With the publication of a new issue of the *Annals* on 1 September, Darwin must have despaired. There was yet another article by the indefatigable Wallace. This time, he had not been satisfied simply to speculate about the orang-utan and the origins of man. He had used his wide experience of the Amazon basin and the Malay Archipelago to offer a radical essay dealing with the process of extinction and divergence among species of birds, a phenomenon which, by extrapolation, could be applied to all kinds of organic beings, including man.[13]

The article took the generalities of the Sarawak Law and made them specific, using observable species. It was a million miles from static discussions about theories of multiple creations, or of migration between continents and oceanic islands.

CHAPTER 14

Wallace and divergence

THE RUSTLING in the trees above his head attracted Wallace's attention as he scoured the forest floor for beetles. It was March 1855, one month after he had mailed the Sarawak Law paper, and one week since he had shifted his focus back to his everyday job of searching for specimens.

High above him, he saw for the first time an orang-utan. He recorded in his species notebook: '[I] saw a large red haired animal moving slowly along hanging from the branches by its arms. It passed in this manner from tree to tree till it disappeared in the jungle which was so swampy that I could not follow it.'[1]

The word 'swampy' underplays how difficult was the terrain in which Wallace was working. In a contemporary letter to the *Zoologist*, he explained:

> the country all round us is dead level and a perfect swamp, the soil being vegetable mud, quite soft, and two or three feet deep, or perhaps much more. In such a jungle it is impossible to walk; a temporary path has however been made from the river (about a mile and a half) by laying down trunks of trees longitudinally. Along this path is very good collecting-ground, but many fine insects are daily lost, and butterflies can hardly be captured at all, from the impossibility of stepping out of the path, and the necessity of caution in one's movements to preserve balance and prevent slipping, not at all compatible with the capture of active tropical insects.[2]

Wallace had gone to Borneo partly because he had wanted to investigate the natural history of the orang-utan, and particularly 'to determine definitely whether or not three species exist here, and also to learn something of their habits in a state of nature'. Over the following three months, Wallace studied orang-utans at every opportunity. He soon convinced himself that there were only two species, rather than three, as reported by others before him. He also discovered that its behaviour when attacked was unlike that described by Lyell in his *Principles*.

Lyell had said that when the orang-utan takes flight from passing danger, 'he immediately falls down on all-fours, showing clearly that this was the original position of this animal'.[3] Wallace, however, observed that when the animal is pursued or attacked:

> his object is to get to the loftiest tree near; he then climbs rapidly to the higher branches, breaking off quantities of the smaller boughs, apparently for the purpose of frightening his pursuers. Temminck [an earlier naturalist] denies that the orang breaks the branches to throw down when pursued; but I have myself several times observed it. It is true he does not throw them at a person, but casts them down vertically; for it is evident that a bough cannot be thrown to any distance from the top of a lofty tree. In one case, a female Mias, on a durian tree, kept up for at least ten minutes a continuous shower of branches and of the heavy spined fruits, as large as 32-pounders, which most effectually kept us clear of the tree she was on. She could be seen breaking them off and throwing them down with every appearance of rage, uttering at intervals a loud pumping grunt and evidently meaning mischief.[4]

One of Wallace's aims had been to address the possibility that man and the orang-utan might both have descended from some common ancestor, now extinct. Wallace found he could not help speculating on these strange creatures, which 'so closely approach us in structure and yet differ so widely from us in many points of their external form'. With his observer's eye now attuned to geographical distribution, Wallace noted that the orang-utan was confined to the two islands of Borneo and Celebes:

> It is a remarkable circumstance, that an animal so large, so peculiar, and of such a high type of form as the orang-Utan, should yet be confined to such a limited district, - to two islands, and those almost at the limits of the range of the higher mammalian; for, eastward of Borneo and Celebes, the Quadrumana and most of the higher mammalian almost disappear. One cannot help speculating on a former condition of this part of the world which should give a wider range to these strange creatures, which at once resemble and mock the 'human form divine'... When we consider that almost all other animals have in previous ages been represented by allied, yet distinct forms, - that the bears and tigers, the deer, the horses, and the cattle of the tertiary period were distinct from those which now exist, with

> what intense interest, with what anxious expectation must we look forward to the time when the progress of civilization in those hitherto wild countries may lay open the monuments of a former world, and enable us to ascertain approximately the period when the present species of Orangs first made their appearance, and perhaps prove the former existence of allied species still more gigantic in their dimensions and more or less human in their form and structure! Some such discoveries we may not unreasonably anticipate, after the wonders that geology has already made known to us. Animals the most isolated in existing nature have been shown to be but the last of a series of allied species which have lived and died upon the earth. Every class and every order has furnished some examples, from which we may conclude, that all isolations in nature are apparent only and that whether we discover their remains or no, every animal now existing has had its representatives in past geological epochs.

Late in 1855, as he drafted the last of three articles on the orang-utan to be published in the *Annals* the following summer, Wallace's most important deduction from his observations was the certainty that man had evolved from some antecedent, more ape-like species (or, possibly, that an ape had evolved from a more man-like one).[5] This declaration of unqualified certainty, based on watching and examining species in the wild, clearly indicated how Wallace's ideas had developed since sending off the Sarawak Law earlier that year.

Such theorising, however, did not pay the bills. Between March and November 1855, he travelled up Borneo's Sadong River, crossed the watershed and made his way back down the Sarawak River in search of rare and interesting species for his customers. In a nine-month trip, he collected ten thousand specimens.[6]

Perhaps the most rewarding period was spent at a cottage in a hilly region twenty miles from Kuching. There, on dark wet nights with a hurricane lamp suspended on a whitewashed balcony, he captured specimens from thirteen hundred species, all drawn to the light. On one particularly fruitful night, he took two hundred specimens representing a hundred and thirty species.

Practically and logistically, such a haul created its own problems of storage, transportation for the remainder of the expedition, and the temporary identification of the time and place of discovery. When he arrived back in Sarawak he faced weeks of work, mounting and presenting and packing the specimens ready for their despatch to England.

During this busy period, Wallace also found time to test the truth of his Sarawak Law, and his belief that if enough examples from the fossil record could be discovered, 'the great gaps that exist between fishes, reptiles, birds and mammals would then, no doubt, be softened down by intermediate groups and the whole history of the organic world would be seen to be an unbroken and harmonious system'.[7]

In May 1855, two months after his encounter with the orang-utan, Wallace began concentrating on the physical characteristics of hornbills, birds he had first observed and collected in Malacca shortly after he arrived in the Malay Archipelago the previous summer. He had previously thought of the bird as uninteresting, and yet, within six months, it was to present him with the key to a clear understanding of how divergence, modification and extinction had operated to produce related species on continents that were ten thousand miles apart. As he skinned and dissected the carcasses of hornbills taken on Borneo, and exposed the bone structure of the birds' feet, Wallace must have realised that this was not the first time he had seen this particular arrangement of bones.

In his book *Narrative of Travels on the Amazon and Rio Negro*, published in 1853, Wallace had written about the habits of birds he had seen there.

> It must strike everyone that the numbers of birds and insects of different groups, having scarcely any resemblance to each other, which yet feed on the same food and inhabit the same localities, cannot have been so differently constructed and adorned for that purpose alone.

Wallace then pointed out that the goatsuckers, swallows, tyrant flycatchers, jacamars, trogons and humming birds he had observed in the Amazon basin all feed on the same kind of food, and procure it in the same manner: they all capture insects on the wing, and then return to a convenient branch or to the same spot on the ground to devour them. He pointed to the difference in 'structure and appearance' of all these birds, but he had been particularly struck by the way naturalists before him had forced species of birds into tribes according to their appearance – their apparent affinity – rather than their shared underlying characteristics.

As an example, he had pointed out that families of hummingbirds and sunbirds had been placed together because a mere outward resemblance had been mistaken for an affinity:

> A similarity in size, in the prevalence of metallic colours, and in the slenderness of a very variable bill, has been taken to over-balance the most important structural differences... The universal aspects of the Hummers are, excessively long wings and as excessively small feet, with more or less united toes. They take their food exclusively on the wing. Every motion is

> made upon the wing. The feet are solely used as means of support, never for locomotion.

Sunbirds, on the other hand, are constructed entirely differently. Unlike hummingbirds, their legs and toes allow them to hop and perch. Wallace was quite convinced that there was no general agreement of structure to unite the two, except for the solitary (and, he felt, trivial) matter of an elongated and slender bill.

When, six years later, he examined the structure of the bones in the feet of hornbills, Wallace realised that there was a direct and immediate relationship between them and the hummingbirds of the Americas. He concluded that the hornbills' comparatively short legs, united toes and broad flat soles meant that they were not only related to hummingbirds, but also to species of other families with similar structural characteristics.

Controversially, Wallace proposed that the hornbills and hummingbirds – among the largest and smallest of land birds, respectively – were highly diverged members of a perching tribe of birds with which neither had before been customarily associated, such as swallows, goatsuckers, trogons, jacamars and kingfishers. Moreover, he set out to show that all these species of birds could be arranged on a direct line between the hummingbirds and the hornbills, in such a way that they represented the only remaining species of perching birds from a sequence that had once included intermediate species which had since become extinct.

Wallace had also observed that the physical characteristics of the hornbills resulted not only in the modification of their feeding habits, but also in leftover fruit on the trees from which they fed. This enabled other birds to exist alongside the hornbills without friction:

> They cannot ... fly quickly from branch to branch, picking a fruit here and a fruit there; neither have they the strength or agility enough to venture on the more slender branches with the Pigeons and the Barbets, but they alight heavily on a branch of considerable thickness, and then, looking curiously around them, pick off any fruits that may be within their reach and jerk them down their throat. When they have gathered all within their reach, they move sideways along the branch by short jumps, or rather a kind of shuffle, and the smaller species even hop across to other branches when they again gather what is within their reach; and long after they have left a tree, the Barbets and *Eurylaimi* find abundance of food on the slender branches and extreme twigs.

It was a telling example of connected species living alongside each other, living off the same food sources. Later in the article, Wallace proposed that some of the bird species he had observed were likely to be the missing links between whole diverged families of birds that fed while perching, hopping or climbing.

His whole argument was nothing less than a practical demonstration of the theory behind his Sarawak Law: that related species not only emanate from a common ancestor, but originate in the same geographical areas as their ancestors, and can exist and thrive side by side in the same environment.

He knew he was about to upset many ornithologists. He explained: 'In this innovation we are not aware of having any support; yet we think it possible to show good reason for it'. Wallace then re-emphasised his position on the species question for those who might have missed the publication of his Sarawak Law: 'it is an article of our zoological faith, that all gaps between species, genera, or larger groups are the result of the extinction of species during former epochs of the world's history'.[8]

In a later passage, Wallace indicated his confidence in the geological record, as he had done in his Sarawak Law:

> Geological investigations prove that the animals now existing on earth are probably not one-tenth, perhaps not one-hundredth, of those which have existed; for all before the Tertiary epoch were of different species and mostly of different genera and thousands of other genera, families and whole orders must have existed of which we are absolutely in ignorance.

Using the geological evidence, he dismissed the ideas of Swainson, one of the leading theorists of the time, as placing artificial limits on the variety and extent of the origination of species.

Wallace's paper, 'An Attempt at a Natural Arrangement of Birds', was sent from Singapore in February or March of 1856, and appeared in the *Annals* on the first day of September 1856, two months after his final article on the orang-utan. It was twenty three pages of common sense about the affinities of birds, based on their underlying characteristics. He made a simple assumption that the evident gaps between the forms, in terms of both size and appearance, had been caused by the extinction of intermediate organic forms over eons. These extinct species, if known, would complete the entire sequence and eliminate the gaps.

Wallace's publication of two well-argued papers on the species question in the *Annals* in two months contrasted sharply with Darwin's attempts to understand geographical distribution by speculating on whether plants could have been transported to different parts of the world by wind or sea currents, or even by icebergs. The September 1856 issue of the *Annals* could not have made very happy reading for Darwin. His theory of species migration was hardly in the same league as Wallace's articles, which utilised the common characteristics of living species across half the world to reveal how variation, modification, divergence and extinction were all factors in the origination of species.

Charles Darwin's essay was by then twelve years old, but still no one except Hooker was fully aware of his theoretical ideas. Darwin had never really opened up to anyone else, not even to Lyell. Hooker kept his doubts to himself, and other colleagues and friends knew only that Darwin was attempting to overcome some remaining difficulties.

CHAPTER 15

A black box in Cambridge

DOV OSPOVAT, a Harvard scholar, was curious to see how Darwin had reacted to the important changes that had occurred in natural history during the years he was devising his theory. In 1974, he was directed towards a black box in the Darwin archives at the University Library in Cambridge. He found that it held a miscellaneous collection of papers and notes that stretched across the whole of Darwin's life. Ospovat was struck by the vast quantity of material that remained, almost entirely ignored by Darwin scholars, from the period 1838–1859: 'The significance of this material became apparent as I arranged my transcriptions of it in chronological order and examined it in conjunction with Darwin's transmutation notebooks and his two essays written in the 1840s.'[1]

On his first visit, in the mid-1970s, Ospovat read only a few of the notes in the black box. He returned in the autumn of 1977, however, with more time at his disposal, and within a relatively short space of time (his book was published only three years later) he made the most sensational discovery. 'Natural selection', as Darwin had first used it in his essay of 1844, was an entirely different idea to 'survival of the fittest', which brought him fame after its use in *On the Origin of Species*. As far as Ospovat could tell, they bore no relation to each other.

For those who had been taught that the Darwinian concept of natural selection as described in the *Origin* had been known to him – and had formed the central idea around which he had built his theory – since 1837, it was a devastating insight. Despite what Ospovat and millions of others had been led to believe by historians and Darwin biographers for more than a hundred years, Darwin's claim (made in his pursuit of priority) that he had understood natural selection for almost twenty years before Alfred Russel Wallace was false.[2]

Ospovat spelled out the difference so that it could not be misunderstood:

> In the essay of 1844, natural selection is not, as it would become in the *Origin of Species*, an ongoing process, working constantly at the improvement of organisms. In the essay it operates only as an organic response to changed conditions.

If Darwin had not understood the true nature of natural selection in 1844, then what else had he not understood? The paper trail in the black box brought Ospovat to the realisation that Darwin not only had no convincing idea in his essay of 1844 of how or why species originate, but he also lacked any understanding of divergence with modification – the central driving mechanism of organic change – until sometime in the latter half of 1856, only three years before *On the Origin* was published.[3]

Ospovat also discovered that until well into the 1850s, Darwin continued to believe that species were perfectly adapted to their environment, and became extinct or exhibited variations only when geological forces changed the world in which they existed.[4]

Until he made this discovery, Ospovat, like most other historians of science, thought that Darwin had moved on from such simplistic ideas much earlier. Certainly, while writing his transmutation notebooks shortly after returning from the *Beagle* voyage, Darwin had believed, as had Lyell, in the idea that the adaptation of organisms to their environment is perfect, that nature is a well-adjusted mechanism, that there is a harmony among organisms and between them, and that the laws of nature were established by God to achieve His ends. In constructing theories of transmutation in the period 1837–1838, Darwin took for granted that variation in nature was extremely rare, except when organisms were isolated in newly formed geological areas and needed to accommodate to environmental change.[5]

For Ospovat, the realisation that Darwin had held onto such ideas until well into the mid-1850s was totally unexpected. Such assumptions, deeply embedded in the traditional view, 'gave [his] theory of natural selection a particular structure, the structure of a mechanism of adjustment to change, a means by which the balance of nature is preserved'.[6]

There was also compelling evidence to indicate that Darwin had not changed his mind after reading Malthus in 1838, when he claimed the idea of natural selection had first occurred to him. If Malthus had influenced him, why were his theoretical ideas still based on Lyell's theory of perfect adaptation until well after he returned to the species question in the mid-1850s?[7]

Ospovat began to realise that Darwin's theoretical position had undergone an unrecorded change at some point between 1844 and 1859, and concluded that the proof of the idea of relative adaptation to changing environmental conditions rather than perfect adaptation was a product of the 1850s, and was not to be found in the notebooks, the sketch or the essay of 1844.[8]

Aware of what he was implying, Ospovat chose his words carefully:

> I do not go so far as to say that Darwin adopted a wholly new theory after 1844, but entertaining that possibility is a useful aid in gaining an appreciation of the magnitude of the changes that occurred in his ideas about evolution and natural selection.[9]

Ospovat was convinced that in 1844, the structure of Darwin's theory was to a large extent determined by assumptions that Darwin had held since before opening his first transmutation notebook. The transformation of the theory that occurred sometime in the 1850s eliminated some of these assumptions, and at the same time introduced some of the most characteristic ideas that are today associated with the theory of natural selection. The idea of perfect adaptation played no role in this new conception of his theory, and instead of a mechanism of adjustment to external change, in order to preserve balance and harmony, natural selection became a force that led inevitably to the progressive development of life.

Somehow, Darwin's understanding of the importance of natural selection had changed. This set Ospovat thinking. If Darwin's theory had changed so drastically, when did Darwin first come near to understanding that there might be some force other than natural selection that caused organisms to continually move away from the original type? When did Darwin first realise that divergence was a crucial element of the evolutionary process?

The question went to the heart of how Darwin became confident enough to write *On the Origin of Species*, and when Ospovat found a note in the Cambridge archive indicating that Darwin had suddenly recognised a system for the division of labour in the animal world, he was convinced it was the moment that elements of Darwin's theory had begun to fall into place after years of thought.

The date of that note seemingly meant nothing to Ospovat, but it should have done. It was written only three weeks after the publication of Wallace's paper on the classification of birds, and the concepts on which it drew belonged to Alfred Russel Wallace.

CHAPTER 16

Crossroads

JUST BEFORE the publication of Wallace's article on the classification of birds, Lyell and Hooker had given Darwin cause to doubt the veracity of his longstanding migration theory, but he was still attempting to convince his friends and colleagues of the weaknesses of Forbes's idea of continental extensions. Now, Wallace's deduction of modification and divergence as a result of comparing the feet of birds touched a nerve.

Eight days after the publication, Darwin wrote about his own studies of the comparative structure of skeletons of species to Hooker:

> I have been working away as usual (floating plants in salt-water inter alia & confound them, they all sink pretty soon, but at *very* different rates) working hard at Pigeons &c &c By the way I have been astonished at differences in skeletons of domestic Rabbits: I showed some of the points to Waterhouse & asked him whether he could pretend that they were not as great as between species, & he answered 'they are a great deal more'.

Ignoring completely Wallace's article, he tells Hooker: 'How very odd it is that no zoologist should ever have thought it worth while to look to the real structure of varieties.'[1]

Then, three weeks after Wallace's discussion of affinities, modification, descent from a common ancestor and geographical distribution in stable and unchanging environments, Darwin wrote himself an aide memoire.[2] It noted:

> the advantage in each group becoming as different as possible, may be compared to the fact that by division of ~~land~~ [sic] labour most people can be supported in each country – Not only do the individuals of each group strive one against the other, but each group itself with all its members, some more numerous, some less, are struggling against all other groups, as indeed follows from each individual struggling.

Five days after writing that note, he wrote to Hooker with an obvious concern. Darwin refers Hooker to a section of the 'Natural Selection' manuscript he is writing. He tells Hooker that he has never felt such difficulty in deciding what to do about geographical distribution. 'It is of infinite importance to me for you to see it, for never in my life have I felt such difficulty what to do' he tells his friend, 'and I heartily wish I could slur the whole subject over.'[3]

The following day, Darwin wrote to James Dwight Dana at Yale University, an authority on volcanic cones and coral structures, declaring that he was sceptical about the immutability of species, but found great difficulty in proving that species could change.[4] He told Dana:

> I have of late been chiefly at work on domestic animals, & have now got a considerable collection of skeletons: I am surprised how little this subject has been attended to: I find very grave differences in the skeletons for instance of domestic rabbits, which I think have all certainly descended from one parent wild stock.... I know that you are not a believer in the doctrine of single points of creation, in which doctrine I am strongly inclined to believe, from *general* arguments; but when one goes into detail there are certainly frightful difficulties... no one subject gives me such trouble as to account for the presence of the same species of terrestrial productions on oceanic islands; for I cannot swallow the prevalent fashion in England of believing that all islands within recent times have been connected with some continent.

A few days later, Darwin asked his cousin, William Fox, for help. He told Fox that he was finding 'the most remarkable differences in skeletons of Rabbits', and wanted to know whether Fox had ever kept any odd breeds of rabbits, and whether he could furnish Darwin with any details.

It is not too difficult to imagine Darwin being desperate to find his own examples of unlikely connections in the skeletal structure of various species of rabbit, attempting to echo in a domestic area of research the incredible variety of forms Wallace is writing about in the wild. He confessed to Fox that he could not keep on going in the same vein; the task he had set himself was defeating him:

> I remember you protested against Lyell's advice of writing a sketch of my species doctrines; well when I begun, I found it such unsatisfactory work that I have desisted and am now drawing up my work as perfect as my materials of 19 years collecting suffice, but do not intend to stop to perfect any line of investigation beyond current work. Thus far and no farther I

> shall follow Lyell's urgent advice. Your remarks weighed with me considerably. I find to my sorrow it will run to quite a big Book.[5]

Darwin still could not account for the exact relationship between varieties and new species and, other than those in Wallace's article, had no examples of such seemingly different species in nature that had modified and descended from a common ancestor. He certainly knew from his research into barnacles that variations occurred in a state of nature without the need for great geological change, but the importance of this discovery seemed to have passed him by.[6] It was not an idea he exploited in the years that followed. He stuck with his theory of perfect adaptation, possibly believing that if he allowed any modification of his central idea, he would fail to expose Forbes as misguided and his own theory, in the process, might be exposed as inadequate.

By mid-October, still seeking help with aspects of geographical distribution, Darwin turned to the botanist Asa Gray in Boston. He expressed surprise to Gray at some affinities between flora found in both the United States and Europe, and asked whether Gray could offer any explanation.[7] On 4 November, Gray wrote back discussing the possible routes by which various plants could have reached the United States from Europe, and suggesting that the most likely of these was overland migration by way of Greenland.[8] Five days later, Hooker replied positively to Darwin's 'Natural Selection' manuscript, but objected that the migration theory failed to account for some of the most interesting sharing of species between mainlands and islands.[9]

It was only ten weeks since Wallace's article on birds, with its discussion of the geographical distribution of species, had appeared in the *Annals*, but Darwin was already seeking more pragmatic explanations than the accidental migration of plants and animals.

On 11 November, he told Hooker that botanical evidence regarding New Zealand and Raoul Island in the Pacific that Hooker had pointed out 'looks more like a case of continuous land, or perhaps of several intervening, now lost, islands than any I have yet seen'. Darwin added that Hooker's verdict on the manuscript he had sent him had been a great relief. He told Hooker, 'I had become so bewildered with conflicting facts, evidence, reasoning and opinions that I felt to myself that I had lost all judgment. – Your general verdict is *incomparably* more favourable than I had anticipated.' He added that it would be months before he came again to the problem of geographical distribution.[10]

However, in the next ten days things begin to move very quickly for Darwin. First, he told Hooker: 'species do become changed and ... time is a *most* important element (which I think I shall be able to show very clearly is the case) in such change'.[11] A few days later, his ideas had changed even more radically.

> The conclusion which I have come to, quite independently of geographical distribution, is that external conditions (to which naturalists so often appeal) do by themselves *very little*. How much they do is the point of all others on which I feel myself very weak.— I judge from facts of variation under domestication, & I may yet get more light. But at present, after drawing up a rough copy on this subject, my conclusion is that external conditions do *extremely* little, except in causing mere variability ... [which] I look at as *very* different from the formation of a marked variety or new species.[12]

Suddenly, external conditions – formerly pivotal in his theory of how species originate – were no longer important to Darwin. Somehow, he had come to believe that changes in external conditions caused variability alone, but that the formation of new species was due to selection. 'The relation of an organism to its associates is far more important than external conditions', he explained to Hooker, in a sentence that is almost a précis of the facts presented by Wallace in his birds article.

Then, in December 1856, Charles Darwin wrote another note to himself:

> What are called important parts vary seldom – and so they differ only seldom in great groups – now it is probable that most diverse are apt to be propagated; if so it would ensure that we should have type with important differences [that is, subtypes].[13]

When he read it, Dov Ospovat knew that that short note had been triggered by something. It had not come from anything significant written to Darwin in letters from his regular correspondents, nor was there a hint of it in anything he had written to his friends, or recorded in 'Natural Selection'. Ospovat never recognised the connection, but the note echoed precisely the central idea in Wallace's birds paper, published exactly three months before. Where Darwin had written 'what are called important parts vary seldom', Wallace had drawn attention to the characteristic physical structure and shape of birds' feet. The second part of Darwin's note – 'probable that the most diverse are apt to be propagated; if so it would ensure we should have type with important differences [i.e. subtypes]' – suggests that in the wake of Wallace's articles, he is at last seeing the process of divergence of different species from a common ancestral species over time. It is surely reasonable to suggest that Darwin had been influenced by Wallace's discussion of modification, divergence and extinction.

However, Darwin at that time was still convinced that transmutation could occur only once species had migrated to previously unoccupied environments newly created by geological change. He still could not accept that 'the same species of terrestrial production on oceanic islands' had come about not by species migration, but by the subsidence of land that had once joined such islands to a nearby mainland. Yet his friends kept offering evidence that his theory of migration simply could not accommodate. In a letter to the Reverend John Stevens Henslow (who had been his geology tutor at Cambridge) sometime after 6 December, Darwin admitted that he still found the subject of varieties 'deeply interesting but horribly perplexed'.[14]

There can be no escaping the conclusion that a week or more into December 1856, Charles Darwin had still not found an answer to the relationship between varieties and new species, and that other crucial aspects of the species problem were still causing him great consternation.

CHAPTER 17

Out of the blue

ONCE HE had sent off his articles on orang-utans and birds to the *Annals* early in 1856, Wallace waited at Singapore for a boat that never came. His aim was to get to Celebes (now Sulawesi), the island between his recent base of Borneo and New Guinea. Boats sailing directly to Macassar (now Makassar), the capital of Celebes, from Singapore were rare, and the months passed slowly.

At that time, Celebes was the least known of the large islands of the archipelago, and Wallace believed that because of its position (at the centre of fourteen thousand other islands sprinkled over an area roughly the size of North America), it could turn out to be a wonderland of unknown species.

Wallace had already spent two years in the Malay Archipelago, but had still to achieve his main objective, which was to investigate the less well-known eastern islands. He was familiar with the Malayan language and had become acquainted with 'the manner, customs and prejudices of the people'. His assessment of his progress up to that point was that he had learned much by experience, and had obtained 'such a knowledge of the productions of the western portion of the archipelago as will add greatly to my pleasure and interest when exploring the eastern'.[1]

The loneliness of his quest for an answer to the puzzle of how species originate was evident:

> I look forward, in fact, with unmixed satisfaction to my visit to the rich and almost unexplored Spice Islands – the land of the Lories, the cockatoos and the birds of paradise, the country of tortoise-shell and pearls, and beautiful shells and rare insects... The physical privations which must be endured during such journeys are of little importance, except as injuring health and incapacitating from active exertion. Intellectual wants are much more trying; the absence of intimate friends, the craving for intellectual and congenial society, make themselves severely felt, and would be

> unbearable were it not for the constant enjoyment and ever-varying interest of a collector's life and the pleasures of looking forward to a time when the stores now amassed will furnish inexhaustible food for study and reflection, and call back to memory the strange, beautiful scenes among which they have been obtained.[2]

When he wrote this summary, he might have expected to have received a certain amount of intellectual inspiration from the gentlemen natural philosophers of London. It was more than a year since he had forwarded his paper on the Sarawak Law to the *Annals*. He also needed some good fortune to revive interest in the specimens he was sending back to London. From Stevens, he had learned that collectors in England were disappointed with his produce. Wallace complained that:

> some persons who have seen that portion of my collections which has already arrived in Europe have been much disappointed, and have complained (almost as if I made the insects as well as collected them) that ... beetles from the North of China, though from a comparatively cold climate, were much finer.[3]

Wallace was aware that collectors and enthusiasts back in Europe had been given a completely false idea of the species and varieties to be found in the Malay Archipelago. Before he had arrived, European collectors living there had bought (from natives and traders alike) specimens captured from vast distances away. Wallace, who was as interested in precisely when and where a specimen had been taken as he was in the specimen itself, felt that this practice was totally unscientific, and misleading to anyone interested in studying animal life.

He had discovered that many of the designated localities of specimens written about by the early French naturalists such as Boisduval, Lesson and Guérin-Méneville were quite inaccurate. His problem was that most collectors at that time believed that God played a role in deciding where new species should be introduced. Since there was no way of knowing where those places were, the scientific importance of precise geographical location was lost on them.

Stalled in his endeavours, desperate for intellectual stimulation and frustrated by amateurs, the waiting in Singapore must have been difficult for someone of Wallace's energy and ambition; but this period proved to be a turning point in his scientific understanding of species and their geographical boundaries. Had the shipping timetable been different or had he managed to catch the last available boat to Macassar, which had left Singapore the day before he arrived at the port, Wallace would have missed one of his greatest practical discoveries.

In mid-May 1856, Wallace gave up waiting for a boat to take him directly to Macassar, and took passage instead for the island of Bali, where he thought he had more chance of finding a boat heading for Celebes. He stayed on Bali for only two days because the cultivated nature of the countryside was not ideal for collecting rare insects and birds. He crossed the fifteen-mile-wide channel to the neighbouring island of Lombok, only to find exactly the same kind of development near the port of Ampenam, where he hoped to find a ship leaving for Macassar.

One day, having a few hours spare, Wallace hired an outrigger to take him to the southern part of Ampenam Bay, where the land was covered by scrubby, thorny vegetation. It was an unlikely place for what was to prove one of the most amazing natural boundaries on earth: a natural boundary between the families, genera and species of the Australian and Asian regions. To the east, the chain of islands leading towards New Guinea exhibited a relative absence of Asian species; to the west, towards Bali and Sumatra, he knew no Australian species. The boundary is still known as the Wallace Line.

In his journal he noted, 'Birds very interesting, Australian forms appear. These do not pass further West to Bali and Java and many Javanese birds are found in Bali but do not reach here'.[4] In a letter to Stevens towards the end of his time on Lombok, he went into some detail about his discovery. He told his agent that 'many other species illustrate the same fact, and I am preparing a short account of them for publication'.[5]

When he first undertook his exotic collecting, Wallace's intent was to study the distribution of closely related species for evidence that might 'elucidate the circumstances in which new species arise', but the distribution pattern he had stumbled upon was quite unexpected. He had been searching for 'fine-scale patterns'. What he found was discontinuity on a much greater scale, for the distribution of entire families was demarcated.[6]

Wallace's need to identify the circumstances in which new species arise had taken him from the furthest reaches of the tributaries of the Amazon towards the outer boundaries of the Malay Archipelago. His discovery of the Wallace Line gave him the chance to offer radical new ideas on how species originate on oceanic islands. It also provided more evidence for Edward Forbes's Atlantis theory, indicating as it did that most oceanic islands were remnants of a former continental mass that had been split up and cut off by incredible geological forces over immense stretches of time. Bali, which shared families, genera and species with the Malay Peninsula and the islands of Sumatra, Java and Borneo, must once have been connected by dry land to the continental land mass of India and China. Lombok, with its entirely different organic forms, must once have been connected to the continental mass of Australia. On both land masses, families of animals had evolved in different ways over hundreds of millions of years.

After ten weeks of noting the organic forms of the Australian zoological province on Lombok, Wallace eventually found a boat to take him to Macassar on Celebes. He arrived there in late August 1856, just as his birds article was being printed in London. Celebes was a disappointment, and the collecting was poor. However, Wallace was not the kind of man to sit around waiting for something to happen. He needed to tell someone about his discovery. His letter to Stevens, with a very brief account of the Wallace Line, was dated 21 August and left the island on the mail steamer which arrived at the end of the month. On 3 November, Stevens read it out to members of the Entomological Society at their monthly meeting in London. In the minutes of that meeting, there is no report of any discussion about the significance or otherwise of Wallace's discovery.[7] Had he known that his discovery had been aired in public, Wallace might have been surprised. As far as he was concerned, he had shared his information only with Stevens.

The discovery's implications for the understanding of the geographical distribution of species, however, meant that he was obliged to spell it out to someone who would be critically aware of its theoretical importance, and to do that he would need to explain where his latest ideas on species had taken him. But where should that information go? Who should be told? Where could he send it to ensure the most interested reaction? It is likely that he was still hurt by the fact that there had been absolutely no response to the ideas that formed the basis of the Sarawak Law. This time, perhaps, a learned journal might not be the best way.

Wallace knew about similar discoveries in the past. There had not been many; the closest example was, however, obvious, and its discoverer would surely be interested in hearing Wallace's news.

In the 1845 version of his *Beagle* journal, Darwin had written about North and South America, 'I know of no other instance where we can almost mark the period and manner of the splitting up of one great region into two well characterised zoological provinces.'

Wallace knew the 1845 version well and had a copy of it with him in Macassar. Now he had another example for Darwin to consider, perhaps an even more significant example than that of North and South America. The first letter Wallace ever addressed to Charles Darwin at his home in

Kent, written on 10 October 1856, contained more information about Wallace's original and radical views on species than Darwin could ever have expected. (Unfortunately, as the letter no longer exists we cannot know exactly what it contained.) Wallace knew that Darwin was one of the few men who would clearly understand the import of what he was now saying about species. His belief in the existence of some 'principle' that regulated the formation of species in the natural world was already on record from his time in the Amazon basin. The Sarawak Law had identified that principle, stating that close relatives of any given species tend to be found nearby, both in geographical space and in geological time as fossil evidence, because there is an evolutionary process that generates a diversity of new species from an ancestral species. At the time he had no idea what this evolutionary process might be, but something certainly caused species to diverge. For Wallace, the Sarawak Law was always only 'the announcement of the theory, not its development'.

The second part of his theory was connected to the first. Exactly what determined the production of new species and varieties in nature had long defied explanation, but his years of observation had convinced Wallace that there was no difference between the two. In explaining to Darwin his ideas on varieties and species, it is likely that Wallace outlined this conviction.

One further element of his core beliefs is likely to have been contained in that first letter. Wallace had long observed that in the natural world, different species with structural characteristics inherited from a common ancestral species could exist alongside other species of the same genre without recourse to newly established and isolated territories.

All these ideas were revolutionary and Wallace's innocent letter to Darwin was to prove highly significant. Wallace dated his letter 10 October. In due course, the letter arrived in Singapore on 15 November, from where – along with the other second-class mail for England – it departed on the steamer *Singapore* eight days later. After a delay to the P&O steamer *Ripon* in the Mediterranean and appalling weather between Malta and Southampton that winter, the mail eventually arrived at Southampton on 11 January 1857, a crucial date in the controversy surrounding when and how Charles Darwin arrived at his theory of evolution as propounded in the *Origin of Species*.

CHAPTER 18

Changing course

WORKING through the papers in the black box at Cambridge, Dov Ospovat noticed that as 1856 ended and 1857 began, Darwin's ideas about species variation began to change. They no longer depended on the natural theological structure of the 1844 essay. Darwin was no longer claiming that variation only ever happened in newly formed and isolated environments where perfect adaptation did not yet apply. Instead, he now claimed that variation could occur at any time, with or without changes in external conditions. It was a complete reversal, and Ospovat was puzzled. There was nothing specific he could point to that might have caused this fundamental change in Darwin's ideas. It certainly had not come either from Darwin's study of barnacles or from a weight of collected evidence.[1]

Quite clear about the huge implications for the development of Darwin's theory, Ospovat spelled out his discovery. In 1844, Darwin had believed there could be no species variation in nature without a change in geological conditions. He had stated: 'geological change alone causes adaptation in species'. He was convinced that species could change only when the environment to which they were perfectly adapted changed. Now, he was of the opinion that 'geological forces are not a causal factor in organic change'.[2]

In a letter to Hooker in November 1856, only two months after the publication of Wallace's birds article, Darwin referred to 'notions' he was testing that 'species do become changed and that time is a most important element ... in such change'.[3] Ospovat drew his deductions:

> Together, these shifts in emphasis and the numerous others that could be added to the list produced a substantially new conception of the evolutionary process. It is the Darwinian conception we are familiar with from the *Origin* but it is not a conception that we are justified in reading back into Darwin's earlier writings.[4]

What Darwin was suddenly arguing was that variation in nature was 'an innate property of organisms', something natural and unavoidable in all circumstances. Ospovat was absolutely convinced of the importance of this change in Darwin's thinking. Everything that Darwin had put at the core of his theory before this point was no longer valid. Suddenly, his new thinking on variation comprised the core of a new theory. Darwin was beginning all over again.

Ospovat was convinced that the mainspring of the theory that led to the *Origin* less than two years later was understood by Darwin only in the last few months of 1856. He was also convinced that what had turned Darwin's theory of natural selection into a theory of progressive development was his discovery of something he called his 'principle of divergence'.[5] There is, however, nothing in Darwin's autobiography, in any of his letters, nor anywhere in his journal, to give an account of how or when he was led to or stumbled upon such a principle.

Years after the *Origin* had been published, when he was celebrated the world over as the greatest naturalist who had ever lived, Darwin described the moment he understood the idea of divergence: 'I can remember the very spot on the road whilst in my carriage when to my joy the solution occurred to me and this was long after I came to Down'.[6] However, of the intellectual process, and of how all his work and experience fused together to illuminate the way ahead, there was not even a hint.

Back in the twentieth century, one researcher had her own reasons for examining the correspondence of Charles Darwin. What she found, or rather did not find, caused her to suspect that things might not have been as open or as transparent between Darwin and Alfred Russel Wallace as everyone had been led to understand.

CHAPTER 19

Most unlucky man

BARBARA BEDDALL, newly armed with a master's degree in zoology from Yale, took exception to Loren Eiseley's suggestion that Charles Darwin had plagiarised the ideas of Edward Blyth in the development of his theory of evolution. Beddall felt that Eiseley must be misguided, and set out to research Darwin's original papers and correspondence in Cambridge in an attempt to prove him wrong.

Beddall was unusual. She had taken her first degree in zoology in 1941, but after eight years as a research librarian at *Time* magazine and seven years as a writer, she returned to zoology and the question of how Darwin had arrived at his ideas. By this time she had gained the reputation of being a formidable researcher.

Starting at the basics, Beddall began her wider research with a decision to concentrate only on primary source material. She knew from Francis Darwin's introduction to his father's letters that there were some gaps in the extant correspondence, but when she began she could have had no idea how crucial those gaps would prove to be. Sifting through the voluminous correspondence, which at that date had not been assembled in date order, Beddall desperately wanted to find the letters exchanged between Wallace and Darwin. They would, she hoped, indicate exactly what Wallace had told Darwin about his own work in that first letter. Instead, she discovered that the letters Darwin had received from Wallace in 1857 and 1858 no longer existed; nor could she find equally important letters from Lyell and Hooker to Darwin in the summer of 1858. Similarly, a vital letter from the American botanist Asa Gray could not be found.[1]

Beddall felt aggrieved. Reporting her lack of success in finding the letters, she said she had not expected such a thing. She had found it 'very odd' that the most critical correspondence in Darwin's files between 1855 and 1858 was missing. She commented: 'Without these letters, a clear idea of the extent of Wallace's influence on Darwin is beyond academic assessment and the full story impossible to gauge'.[2]

Beddall believed Darwin had been much more aware of Wallace than he had ever let on, and that someone had 'cleaned up the file'.[3] In her opinion, Darwin's son Francis had destroyed the missing letters after his father's death. The idea that it might have been Darwin himself seems not to have occurred to her.

Although the files might have been purged, there was an intriguing anomaly. Beddall discovered that in the entire Darwin archive, just one small scrap remained of a letter Wallace had written to Darwin. It came from a second letter, which Wallace had written to Darwin in September 1857. Of the first, sent from Macassar in October 1856, there was no trace. The absence of the letters was a great disappointment and Beddall was not convinced by Francis Darwin's explanation that when files were full, his father discarded old letters to make room for new.[4]

Darwin had covered his tracks well. He always maintained that he had not received the first letter from Alfred Wallace until the end of April 1857 (rather than in January of the same year).[5] However, the timings do not add up. In order to accept Darwin's assertion, one would have to believe that the most efficient contemporary postal service in the world had taken six months to transfer a letter from the Malay Archipelago to Charles Darwin's home, rather than the promised two months, without any explanation of where it had been in the meantime. Further, one would have to accept that, due to an extraordinary coincidence, in that lost interval of four months, Charles Darwin conceived of an entirely new species theory and turned away from ideas he had been wedded to for the best part of twenty years. Alternatively, one could theorise that Wallace's letter had indeed arrived at Darwin's home at the beginning of the year, and had been in Darwin's possession during those first four crucial months of 1857, during which his ideas underwent such a dramatic change.

CHAPTER 20

The collector's life

ONE IRONY of this story about the origination of one of the most far-reaching scientific theories is that it turns, not on a brilliant eureka moment at Down House, but on the whereabouts of some second-class mail posted from an island in the Malay Archipelago.

The unyielding precision (barring accidents) of the Victorian mail service is evident in records kept by the Post Office and in public announcements of arrivals and departures of vessels carrying the mail from and to outposts of the British Empire. Before the summer of 1857 mail ships from India and China left for Great Britain once every month, but with the outbreak of the Indian Mutiny the British government caused the P&O shipping line to increase the frequency of its service from Calcutta, Bombay and Singapore to Southampton to once every two weeks. News of the desperate military and political situation in India was carried to England by the P&O liner from Calcutta, which picked up the mail from Bombay and Singapore at Galle in Ceylon (now Sri Lanka) and rushed it home via Suez, Alexandria, Malta and Gibraltar.

For Wallace, the solitary naturalist, the complications of the postal schedules would have been a long way from his thoughts as he considered his prospects of profitable collecting on the island of Celebes. It was enough for him to know that if he posted a letter early enough for it to be picked up at the end of that particular month by the Dutch mail steamer on its journey round the archipelago, then that letter would be in London nine weeks later. He had already written to Stevens at the end of August and this letter duly arrived in early November.

The letter he had now begun writing to Samuel Stevens would take a similar route and an almost identical journey time. If he could finish it in time to catch the mail boat, it would leave Celebes at the end of September and be in London with the rest of the second-class post in early December.

After the excitement of Lombok, Wallace found Celebes dull. The island had turned out to be unproductive and boring. In his journal, three weeks after he had arrived, he complained about the nakedness of the island, and the flat landscape that stretched for miles around the capital of Macassar (now Makassar). For half of the year it was covered with water; during the other half it was an expanse of baked mud 'with only an apology for vegetation'.[1]

Using a patch of forest between six and eight miles away from his bamboo house as a collecting ground, he managed to take some birds and butterflies, 'but no beetles'. Miles of cultivated ground, 'barren for the naturalist', extended around every town and village, and to get to the uncultivated areas was both difficult and expensive. In many places there was a risk of being robbed or worse. However, Wallace's natural enthusiasm could not be dimmed for long. He wrote in his latest letter to Samuel Stevens:

> I hope soon to make arrangements for a small house near the forest I have spoken about where I can stay a week at a time, and then bring home and store my collections at my house near Macassar: already I can see that I shall get a pretty good collection of birds. Raptorial birds are abundant (the first place I have seen them so in the Archipelago); I have already seven species, one or two of which I have no doubt are new.[2]

Wallace identified Bontyne, sixty miles north of Macassar, as a place where he hoped to find the best collecting. The overland route would be difficult, given all the luggage the collector needed to take with him; the sea journey, with prevailing winds in the wrong direction, could take up to a month. So he planned to go to Bontyne in January, when the winds would be favourable and the journey would take just 24 hours. Once there, he expected to find many interesting specimens in the forested highlands of nearby mountains.

The letter gave insights into the planning and organisational side of a collector's life. Wallace addressed it to Stevens, dated it Macassar, 27 September 1856, and posted it.

The certainty with which assertions can be made about the efficiency and reliability of the postal service in the nineteenth century may surprise some modern readers but the reality is the system worked exceptionally well and there exists a huge body of archive material to verify this. In addition to shipping records that still survive in postal archives, contemporary newspaper reports published in London, Bombay, Colombo, Singapore and Batavia (modern-day Jakarta) can be pieced together to build up a reliable timetable of which vessel carried which bundle of mail on a specific date. For the specialist field of shipping movements in the Dutch East Indies during this period, much credit goes to Professor Femme Gaastra, professor of Dutch Maritime History at the University of Leiden in the Netherlands, who was instrumental in providing the author with a detailed analysis of the shipping reports from the *Javasche Courant.* Published twice-weekly in Batavia throughout the nineteenth century, the *Courant* was an essential business newspaper that not only gave its readers indispensable commercial information at the time but it has also since proven to be the bedrock of all economic analysis of the region for academic researchers.

Analysis of the Batavia shipping reports by Prof. Gaastra shows that the next ship available to carry Wallace's letter to Samuel Stevens was the steamship *Padang*, which left on 28 September 1856, the day after he posted the letter. The *Padang* in turn transferred its mail to the *Koningin der Nederlanden*, which sailed from Batavia on 11 October[3] and arrived in Singapore on 14 October.[4] Eleven days later it was aboard the P&O steamship *Malta*[5] in closed boxes and supervised by a former Royal Navy officer. The letter had begun the next leg of the journey to Southampton. The last leg of that journey from Alexandria to England was to have been aboard the P&O steamer *Candia* but 'an accident to the machinery of the steamer' meant that the Admiralty agent on board, Lt. T. Tickell, R.N., escorted the mail up through France from the Mediterranean and across the Channel to London, where it eventually arrived, two days late, on 7 December 1856. Sometime in January, Stevens must have handed it to the secretary of the Entomological Society since it was published, along with all other mail received that month, in the society's magazine at the end of February 1857.[6]

About a fortnight after he wrote this latest letter to Stevens, Wallace wrote for the first time to Charles Darwin. Dated 10 October 1856, the letter to Darwin was picked up in Macassar by the *Makassar* on 31 October and subsequently fed into the Batavia-Singapore-Suez-Southampton mail route. Thus, three separate letters were sent by Wallace to London over a two-month period.

The metronomic consistency of the mail service from the Malay Archipelago to London one hundred and fifty years ago, with systems in place to safeguard the mail at every stage of the journey, indicates that letters could be posted with absolute confidence in the knowledge that, acts of God notwithstanding, they would be received safely and on time on the other side of the world.

It is, therefore, ironic that three letters, two of which were relatively mundane and the third scientifically fascinating, sent in August,

September and October from Macassar could endure quite different fates – two arriving safely at the home of Samuel Stevens in London, while the last one, a staggeringly important letter to Charles Darwin, sent from the same location a month after he had written to Stevens and held in a secure mail system until opened in Post Office headquarters in London, should 'disappear' from the system for nearly four months until seemingly turning up at Darwin's home just outside London a few days before the end of April. There were no reports of shipwrecks or train wrecks that could explain such a delay. In fact, the newspapers of the time indicate that the postal system was running the way it should – like clockwork.

When Darwin wrote, in his reply to Wallace on 1 May, that he had received his letter 'a few days ago', his words raise more questions than they answered.[7] At this date, the most recent delivery of second-class mail from Macassar had arrived in Southampton on Friday 3 April.[8] The next second-class delivery of mail from Macassar was aboard the steamer *Colombo*, which did not arrive in Southampton until 2 May.[9]

Allowing for subsequent transfer to the mail train and a brief time in the General Post Office in London, the next realistic date for Darwin to have received Wallace's letter at his house was 3 May – two days *after* Darwin dated his reply. But Darwin was not at his home on 3 May. He was at a water treatment spa at Moor Park in Surrey, on the other side of London. So the letter could not have reached him until 4 May even if it had arrived on the *Colombo*.

For Darwin, in this case, a little knowledge was indeed a dangerous thing. He felt safe in his untrue assertion of the date of his receipt of Wallace's letter, because any first-class mail from the East Indies had, indeed, arrived in London on 27 April, having been sped from Marseilles to London by train and boat.[10] From his continuing correspondence with Edward Blyth in Calcutta, Darwin knew when ships servicing the Far East arrived back in England, but he did not yet know that the impecunious Wallace habitually sent his mail second class.

It was not the only time that Charles Darwin claimed to have received a letter later than it had actually arrived. But in that case the dispute over the delivery date has resulted in a cat-and-mouse game which has continued among biographers for nearly forty years.

CHAPTER 21

Land bridges connect

THE MOST significant thing about the arrival of Wallace's letter was that Darwin mentioned it to no one, even after replying to Wallace four months later. It was to be another sixty years, long after his death, before anyone found out that his correspondence with Wallace had begun with the letter he claimed arrived at his home in April 1857.

From Darwin's reply dated 1 May 1857,[1] we have a clear indication that Wallace had spelled out his latest ideas. Darwin wrote: 'By your letter and even still more by your paper in *Annals*, a year or more ago, I can plainly see that we have thought much alike and to a certain extent have come to similar conclusions.' Yet when Wallace had written his letter, the views of the two men had still been miles apart. No one could have described Darwin's ideas on species as having been 'much alike' those of Wallace.

In November 1853, Wallace had lectured London entomologists on geographical distribution and the spread of species from older geological areas to those more recently formed. In a book in 1854, he had indicated his belief in a principle that governed nature. In September 1855, he had published his Sarawak Law about the divergence of organic forms, and made clear the importance of extinction in creating gaps in the evolutionary record. In his *Annals* article of July 1856, he had suggested the possibility that the orang-utan and man might be related. Two months later, he wrote that species of birds that catch food in mid-air in environments as far apart as New Guinea, Africa and the Americas were closely related, not because they outwardly resembled each other, but because of underlying and unchanging characteristics in physical structure.

On the other hand, in the summer of 1856, before Wallace's birds article was published, Darwin was still fighting a rearguard action against Forbes's idea of land bridges, and continuing to assert that organic change only ever followed geological change. He was still clinging to his conception of a world in which organisms emerged perfectly adapted to the environment in which they were placed, and where change could happen only in newly formed and isolated environments. On New Year's Day 1857, he wrote to dispute Asa Gray's suggestion that the passage of European plants to the United States might have been via Greenland, 'because it "riles" me dreadfully'.[2]

Within weeks, Darwin – who had by now received Wallace's letter – was feeling the pressure of Wallace's published progress on the species question. He wrote to his cousin Fox, with a rare admission of how vulnerable he felt:

> I shall take a little holiday sometime; perhaps to Tenby: though how I can leave all my experiments, I know not. – I am got most deeply interested in my subject; though I wish I could set less value on the bauble fame, either present or posthumous, than I do, but not, I think, to any extreme degree; yet if I know myself, I would work just as hard, though with less gusto, if I knew that my Book would be published for ever anonymously. [3]

At this point, Darwin was about to commit to his species manuscript one of the most fundamental elements of his new ideas, yet, in all his letters, to Hooker, Fox, Lyell and others, there is no hint of the influence of Alfred Russel Wallace.

On 31 March 1857, ten weeks after the arrival of Wallace's letter in London, Darwin completed two handwritten pages in Chapter 6 of his 'Natural Selection' manuscript under the heading 'Extinction'. In the final paragraph of a two-page section, he wrote: 'The principle of divergence, I believe, plays an important part in the affinities or classification of all organic beings'.[4]

This was the first indication anywhere in Darwin's work that he had stumbled upon the idea of divergence but there remains no explanation of how it was connected with the theory he had sketched out fifteen years before. In his notes and manuscript for 'Natural Selection', there would be no further reference to or explanation of those nineteen words. That two-page section no longer exists; Darwin excised it from his manuscript in June 1858. Yet we know that on the last day of March 1857, with Wallace's letter in his possession, Charles Darwin claimed that the principle of divergence was an idea that had newly occurred to him.

There is absolutely no suggestion in any of his letters, correspondence or papers of Darwin stumbling upon new ideas (apart from an idea taken from Wallace's birds paper) that might have caused such a radical change in his thinking. Furthermore, once it had been written down, Darwin had dated evidence of when he had first referred to his 'principle of divergence'. Since no one knew of the existence of Wallace's letter, no one could ever prove that it had arrived earlier than Darwin had claimed; nor could anyone speculate that Wallace's announcement that he had understood the importance of divergence might have formed part of that letter.

Even more audaciously, twelve days later Darwin wrote to Hooker:[5]

> I have just been putting my notes together on variations *apparently* due to the immediate and direct action of external causes & I have been struck with one result [...] which seems to me to be most simply explained by species, being only strongly marked varieties & therefore following same laws as recognised and admitted varieties. [...] I have quoted the foregoing remark only generally with no examples, for I add there is so much doubt and dispute what to call varieties; but yet I have stumbled on so many casual remarks on *varieties* of plants on mountains being so characterised that I presume there is some truth in it. What think you?

Wallace could not have explained his species ideas to Darwin without stressing his long-held certainty that he believed species to be only strongly marked varieties. Indeed, this had been Wallace's core belief since reading *Vestiges* thirteen years earlier. Darwin's hesitancy does not suggest this is an idea he is familiar with or has really thought out. Moreover, his comment that 'I presume there is some truth in it' suggests that it is something he has read and quoted, rather than his own original thought.

No other convincing explanation of how Darwin arrived at such crucially important ideas at that particular time has ever been offered. The following day, he wrote to Lyell about Forbes's Atlantis theory with a markedly different tone:

> though as a general rule I am much opposed to the Forbesian continental extensions, I have no objection whatever to its being proved in some cases. Not that I can admit that W[ollaston] has by any means proved it; nor, I think, can anyone else, till we know something of the means of distribution of insects. But the close similarity or identity of the two Faunas is certainly very interesting.[6]

The echoes of Wallace's discoveries are suddenly everywhere in Darwin's work.

On 1 May 1857, as mentioned previously, Darwin wrote to Wallace saying that his letter dated 10 October had been 'received a few days ago':

> By your letter and even still more by your paper in *Annals*, a year or more ago, I can plainly see that we have thought much alike and to a certain

> extent have come to similar conclusions. In regard to the Paper in *Annals*, I agree to the truth of almost every word of your paper; & I daresay that you will agree with me that it is very rare to find oneself agreeing pretty closely with any theoretical paper; for it is lamentable how each man draws his own different conclusions from the very same fact.[7]

When McKinney, the Wallace advocate who was pursuing his doctorate, read this, he was brought up short. He pointed out that according to the notes Darwin had pinned to his December 1855 copy of the *Annals*, he certainly had not agreed with 'almost every word' of Wallace's paper.[8] In fact, Darwin's comments were almost entirely in opposition.

Darwin's letter to Wallace continued:

> This summer will make the 20th (!) year since I opened my first note-book, on the question how and in what way do species and varieties differ from each other. I am now preparing my work for publication, but I find the subject so very large, that though I have written many chapters, I do not suppose I shall go to press for two years… It is really *impossible* to explain my views in the compass of a letter on the causes and means of variation in a state of nature; but I have slowly adopted a distinct and tangible idea. Whether true or false others must judge; for the firmest conviction of the truth of a doctrine by its author, seems, alas, not to be slightest guarantee of truth. … One of the subjects on which I have been experimentising and which cost me much trouble, is the means of distribution of all organic beings found on oceanic islands; and any facts on this subject would be most gratefully received. Land-Molluscs are a great perplexity to me.[9]

The last line indicates that Darwin, at least when he wrote that letter, had still not accepted the superiority of Forbes's land bridge theory to that of organic migration to oceanic islands.

When McKinney studied Darwin's reply, he saw that even though Wallace's letter was missing from the files, one could certainly deduce that Wallace had outlined his ideas on variation and species in his letter. McKinney was aware that Darwin had not before expressed such ideas in his writings or in his journal, and commented that this statement must have aroused Wallace's curiosity, since Darwin was outlining Wallace's own ideas back to him and claiming, in the process, that Wallace's thoughts were his own.[10]

Dov Ospovat was also puzzled, because without any explanation of how he came to understand the principle of divergence, Darwin incorporated it as a new theory into the heart of his species manuscript.[11] Darwin also claimed that his notes now indicated that the production of varieties leads to the production of species.

Ospovat was convinced there was no indication in Darwin's notes that his data indicated any such thing.[12] Moreover in 1980, the Darwin expert Janet Browne said she was sure from her research in the archive in Cambridge that Darwin had not understood the relationship between varieties and species before the summer of 1857.[13] This would have been six months after he received Wallace's letter. So why, today, is Charles Darwin believed to be the man who first had these ideas? The explanation begins with an innocuous letter from the United States.

On 1 June, Darwin received a letter from Asa Gray disagreeing with an idea he had expressed.[14] It was the opportunity Darwin had been waiting for. In his reply two weeks later, Darwin thanked Gray for his remarks, and revealed his current thinking that the process of extinction was responsible for both the wide variety of species and the differences between species. 'I look at Extinction as common cause of *small* genera and *disjoined* ranges & therefore they ought, *if they behaved properly* & as nature does not lie to go together'.[15]

On 7 July, Gray replied. He challenged Darwin to offer convincing evidence of his claims.

> 'It is just such sort of people as I that you have to satisfy and convince, and I am a very good subject for you to operate on, as I have no prejudice nor prepossessions in favor of any theory at all. I never yet saw any good reason for concluding that the several species of a genus must ever have had a common or continuous area. Convince me of that or show me any good grounds for it … and I think you would carry me a good way with you.[16]

Darwin responded with a confession:

> I should like to tell you (and I do not *think* I have), how I view my work. Nineteen years (!) ago it occurred to me that whilst otherwise employed on Nat. Hist. I might perhaps do good if I noted any sort of facts bearing on the question of the origin of species; & this I have since been doing… But as an honest man I must tell you that I have come to the heterodox conclusion that there are no such things as independently created species: that species are only strongly defined varieties. I know that this will make you despise me.- I do not much underrate the many *huge* difficulties on this view, but yet it seems to me to explain too much, otherwise inexplicable, to be false.[17]

Darwin's confidence, now high, allowed him to approach his closer friends with new theoretical propositions. On 1 August, he wrote to Hooker saying that as a result of tabulating zoological data he had met with one of the most important arguments that 'varieties are only small species, - or species only strongly marked varieties'. He added: 'The subject is very important for my work, though I clearly see *many* causes of error'.[18] However, Darwin goes no further, nor does he offer any explanation of how he has suddenly come up with this idea.

In a follow-up letter three weeks later, he indicated to Hooker that he is having made a tabulation of varieties that explains the branching of forms. Then, for the first time, he mentions to Hooker that he has established a principle of divergence, 'which I think I can explain, but which is too long and perhaps you would not care to hear.'[19]

It is obvious that Darwin's understanding of several crucial aspects of the species question had advanced exponentially since the beginning of the New Year. Gone were the stumbling attempts to show seed flotation possibilities or vitriolic diatribes against Forbes's theories. For the first time, Darwin saw things exactly as Wallace had been seeing them for nearly three years.

Darwin must have experienced some trepidation about how Asa Gray would respond to his ideas. He had opened up to the American botanist, taking the risk of outright rejection. When Gray's reply arrived that August, Darwin must have felt great relief. He had discovered a fellow traveller. In his letter, Gray admitted to Darwin that he had many misgivings of his own about the definiteness of species: 'My notions about *varieties* are, I believe, just what you would have them ... and that there is some law, some power inherent in plants generally prompting them to originate varieties'.[20]

Darwin knew that Gray's reply was a significant admission for a man who had long believed that God created all living things. It must have given Darwin a tremendous boost to know that ideas previously advocated only by a fly-catcher in Borneo, and which he was now offering as his own, were readily agreed to by one of the most respected botanists in the world.

Gray had been led to his own confession because of Darwin's newfound confidence and apparent originality. Suddenly Darwin's friends, first Hooker and now Gray, were being allowed to understand that at long last he was developing a theory with which he might well solve the species question. Neither of them would ever learn of Wallace's letter and its influence on Darwin after December 1856, nor was Wallace's name ever used in those crucial letters between Darwin and Gray and Darwin and Hooker.

The brief section in Natural Selection headed 'Extinction', which had been written on the last day of March that year and had remained unaltered at fewer than three hundred words, was the only reference to a principle of divergence in Darwin's archive before these letters were written. His claim to have discovered the principle did not lead Darwin to amend his long-held theory. He did not realise the difficulties he was about to encounter in attempting to fuse together ideas derived from two entirely different theories.

CHAPTER 22

Enter a sceptic

JOHN BROOKS, an environmental biologist at Yale University, was preparing material in the early 1960s for an undergraduate seminar course on evolution. As he became increasingly familiar with the original scientific contributions to the subject from both Wallace and Darwin, he became suspicious. Certain similarities struck him between the concepts, and even the wording, in Wallace's papers and in Darwin's *Origin*. 'Were these really coincidences of two totally independent conceptions? Or did Darwin somehow profit from Wallace's papers – a possibility to which Darwin gave no recognition, not even a hint?'[1]

For Brooks, there were too many similarities. Since most of the troubling similarities centred on the single word 'divergence', Brooks wanted to know more about the chronology of Wallace's discoveries, how much of that new information had found its way back to England before June 1858 and exactly when Darwin first grappled with and then developed his own version of divergence.

To be sure about the exact path Wallace had taken towards his ideas while in the Malay Archipelago, Brooks retraced McKinney's footsteps, studying Wallace's journals and notebooks at the Linnean Library in London.[2] Slowly, he pieced together the dates of Wallace's discoveries, and how and when his theories were transmitted back to London. By the early 1970s, he had a very clear idea of Wallace's journey of discovery, and was absolutely convinced that the tiny islands of Aru, one hundred miles west of New Guinea, had great significance for the development of Wallace's theories.

Brooks knew that Wallace's experiences there in the months between sending off his first letter to Darwin and returning to Macassar the following summer had made a great impression on him. First of all, it had been a very good location for collecting: in just six months, Wallace captured nine thousand specimens representing sixteen hundred species. Wallace was to remember it as 'the portion of my travels to which I look back with the most complete satisfaction'.[3] Second, and more importantly, his stay on the islands had helped him to consolidate his idea that in nature, there is no difference between the appearance of a variety and the appearance of a new species. The distinction, he decided, was false. Wallace's experiences in the islands convinced him that the orthodox species ideas of his generation of natural philosophers in Europe were more than misleading: they were absolutely wrong.

In June 1857, in Macassar, as he pinned and presented his specimens from Aru in their boxes, Wallace must have been rehearsing how and to whom he should offer his evidence and opinions. He decided on three approaches. The first would be the entomologists; the second he decided to send to the zoologists; and the third he would write for the gentlemen naturalists who read the *Annals*. Wallace had decided that this was no time to hold back.

For the Entomological Society, he wrote a brief paper on the habits and markings of a new species of bird-wing butterfly he had recently taken on one of the small islands close to Aru.[4] He indicated that he considered the new species very similar to the specimen from nearby Amboina that Linnaeus had identified in 1758, and also to a specimen taken on neighbouring New Guinea in 1845. Wallace found himself fascinated by the close resemblance of the specimens, but also by the marked differences between them. Linnaeus's Amboina butterfly had four spots on its smaller wings; the butterfly taken on New Guinea almost ninety years later had only two spots. The insect Wallace had captured on the island of Dobbo close to Aru had three.

Just in case the point escaped his audience, Wallace spelled it out. His recent find was an intermediary between the other two. It was the smallest possible difference between two specimens, yet it was, none the less, a difference. The question he posed was simple, but it lay at the heart of the species question: did this differentiation of one spot make his butterfly a variety or a new species?

For the *Zoologist*, he wrote a piece that was extremely direct and amazingly confident.[5] He offered a new way of differentiating varieties and species, dismissing by implication the theories of Charles Lyell, which were shared by many of its readers. Wallace argued that the determining factor in such debates should not be the *degree* of difference, but the *nature* of the difference:

> If permanent characters do not constitute a species when those characters are minute, then a species differs from a variety in degree only, not in nature and no two persons will agree as to the amount of difference neces-

> sary to constitute the one, or the amount of resemblance which must exist to form the other... Now the generally adopted opinion is that species are absolute independent creations, which during their whole existence never vary from one to another, while varieties are not independent creations, but are or have been produced by ordinary generation from a parent species.
>
> There does, therefore (if this definition is true), exist such an absolute and essential difference in the nature of these two things that we are warranted in looking for some other character to distinguish them than one of mere degree, which is necessarily undefinable.

Referring to his example of the number of spots on the wing of a butterfly, he argued:

> If there is no other character, that fact is one of the strongest arguments against the independent creation of species, for why should a special act of creation be required to call into existence an organism differing only in degree from another which has been produced by existing laws?

Wallace then argued specifically and directly against Lyell's notions:

> If an amount of permanent difference represented by any number up to 10, may be produced by the ordinary course of nature, it is surely most illogical to suppose and very hard to believe, that an amount of difference represented by 11 requires a special act of creation to call it into existence.

Wallace concluded:

> To escape this difficulty there is but one way: you must consider every group of individuals presenting permanent characters, however slight, to constitute a species; while those only which are subject to such variation as to make us believe they have descended from a parent species, or that we know have descended, are to be classified as varieties. The two doctrines, of 'permanent varieties' and of 'specially created unvarying species' are inconsistent with each other.

In his article for the *Annals*, he was even more specific.[6] He argued that Lyell's belief that special creations were placed in new locations was simply unacceptable, and that his own ideas about extinction and new species showed them to be so.

As a result of the deep understanding he had experienced on Aru,

Wallace was prepared to take on the world of natural philosophy as few before him had dared to do. Yet Wallace had originally travelled to the island to achieve one of his big ambitions as a collector, rather than as a scientist. In a letter to Samuel Stevens, he wrote:

> Rejoice with me, for I have found what I sought; one grand hope in my visit to Aru is realised: I have got the birds of Paradise (that announcement deserves a line of itself); one is the common species of commerce, the *Paradisea apoda*; all the native specimens I have seen are miserable, and cannot possibly be properly mounted; mine are magnificent… I believe I am the only Englishman who has ever shot and skinned (and ate) birds of Paradise, and the first Englishman who has done so alive, and at his own risk and expense; and I deserve to reap the reward, if any reward is ever to be reaped by the exploring collector.[7]

However, during the last month of his stay on Aru, confined to his house with inflamed legs that were covered with sores caused by insect bites, it was not the bird of paradise that commanded Wallace's thoughts. On arrival, he had expected to find on Aru an impoverished version of the 'immense diversity and richness' of animal species on New Guinea.[8] However, with the singular exception of the bird-wing butterfly he had taken at Dobbo, every species of Australian biota to be found on New Guinea also existed on Aru. As far as he knew, Aru had no species of Asian biota on any of its islands; yet on Ke, an island sixty miles to the northwest but about the same distance as Aru from New Guinea, there were animals of both Australian and Asian derivation. Obviously, the geological history of the islands was different, but how could their difference in species be so great when they were almost equidistant from the western coast of New Guinea? Organic migration could not account for this phenomenon.

A second interesting fact about Aru was that it had two deep, narrow channels running east to west, through 'irregular, undulating, rocky country', separating the island into three segments. They were of uniform width and depth, and closely resembled river valleys. They were filled with salt water, since they were open at both ends to the sea.[9] The third fact of interest was that the birds and mammals of the deep forest of the large, central island of Aru itself, including a kangaroo and several species of smaller marsupials, were all known from the deep forest of New Guinea.

It was obvious to Wallace that the island of Aru, unlike the island of Ke, bore a special relationship to New Guinea.[10] When he found the rocky channels, he began to speculate that Aru and New Guinea had once been joined together. The channels could have been caused by rivers that originated on New Guinea and cut through Aru when the two islands were still connected.

Brooks recognised the insight that had come to Wallace:

> [He] ... had long sought a situation in which he could find evidence of linkage between the formation of new elements in the organic world and recent physiographic change. He had at last found it. One would expect little change following a recent physiographic alteration. According to Lyell, extinctions would occur one by one. Possibly some had occurred; it was impossible to know which of the species that had lived on this terrain when it was a part of a larger New Guinea had subsequently become extinct.
>
> But the one novel element of the fauna was predictable under Wallace's hypothesis; the new form (was it a species or a variety?) was derived from a pre-existing closely allied species. The Aru bird-wing was very like the New Guinean species. Lyell, on the other hand had said that new species were created to fit the new situation, entirely without relation to species that had existed under prior conditions.
>
> Aru thus presented Wallace with a situation that could be used to test publicly his previously published hypothesis as opposed to Lyell's in explaining a set of observable facts.[11]

In 'On the Natural History of the Aru Islands', published in the *Annals* as a supplement late in 1857, Wallace used comparisons with several other islands to show that Aru had once been a part of New Guinea. He bolstered his view by arguing that Aru's rock valleys were of such uniform width and depth that they must once have been true river channels. The separation of the two islands had been caused by the subsidence of the intervening land. It was a classic example of Forbes's land-bridge theory. (As he was on the other side of the world, Wallace could have had no idea of the time Charles Darwin had spent trying to persuade his friends that Forbes's ideas were misguided.)

In support of his discovery, Wallace pointed out:

> families of birds abundant in species and individuals on the Celebes, Sumatra, Java, Borneo and the Philippines are everywhere common birds but not one of all these families of birds was to be found on Aru. Nor, with two doubtful exceptions, were they to be found on New Guinea. The whole are also absent from Australia. In addition not one species of

> mammalia that can be found on any of the great islands westwards from New Guinea and Aru are to be found on those two islands. On Aru and New Guinea all the mammals are marsupials but in the great islands of the western archipelago there is not a single marsupial.[12]

Wallace turned to the theories of contemporary naturalists to see whether they could explain the phenomena of the Aru and New Guinean fauna:

> We know (with a degree of knowledge approaching to certainty) that at a comparatively recent geological period, not one single species of the present organic world was in existence; while all the vertebrata now existing have had their origin still more recently. How do we account for the places where they came into existence? Why are not the same species found in the same climates all over the world? The general explanation given is that as the ancient species became extinct, new ones were created in each country or district, adapted to the physical conditions of the district. Sir C. Lyell who has written more fully, and with more ability, on this subject than most naturalists, adopts this view.[13]

He contrasted the similar climate and physical properties of two islands such as New Guinea and Borneo, where the fauna are entirely different but the climates are similar, with those of Australia and New Guinea, which are entirely different in climate, aridity and size, but where the fauna are strikingly similar. He argued that there must be some reason why the theories of Lyell did not fit these climatic facts. He postulated that New Guinea had once been part of Australia, sharing the same climate. Since they had split apart (and Wallace accepted this as fact more than 70 years before the theory of continental drift was suggested), Australia was predominantly dry while New Guinea became wet and humid; yet it was possible to find kangaroos in the dry deserts of Australia and in the dense rainforests of New Guinea and Aru.

> We can hardly help concluding, therefore, that some other law has regulated the distribution of existing species than the physical conditions of the countries in which they are found, or we would not see countries the most opposite in character with similar productions, while others almost exactly alike as respects climate and general aspect, yet differ totally in their forms of organic life.[14]

Wallace then began to expand upon his Sarawak Law. It was more than two years since he had written it:

> In a former Number of this periodical, we endeavoured to show that the simple law, of every new creation being closely allied to some species already existing in the same country, would explain all these anomalies, if taken in conjunction with the changes of surface and the gradual extinction and introduction of species which are facts proved by geology.

Wallace then claimed, for the first time in public, that there had been a time when islands like New Guinea and Aru had been joined together, as well as being joined to the bigger land mass of Australia. He argued:

> when New Guinea and North Australia were united, it is probable that their physical features and climate were more similar, and that a considerable proportion of the species inhabiting each portion of the country were found over the whole... After the separation took place, we can easily understand how the climate of both might be considerably modified and this might perhaps lead to the extinction of certain species. During the period that has since elapsed, new species have been gradually introduced into each, but in each closely allied to the pre-existing species, many of which were at first common to the two countries. This process would evidently produce the present condition of the two faunas, in which there are many allied species – few identical... It is quite unnecessary to suppose that new species have ever been created 'perfectly dissimilar in forms, habits and organization' from those which have preceded them; neither do 'centres of creation', which have been advocated by some, appear either necessary or accordant with facts, unless we suppose a 'centre' in every island and in every district which possesses a peculiar species.[15]

Wallace had recently received a letter from Henry Bates, who was still working in the Amazon. Bates wrote effusively about Wallace's rate of progress:

> I received about six months ago a copy of your paper in the Annals on 'The Laws which have governed the introduction of New Species'. I was startled at first to see you already ripe for the enunciation of the theory. You can imagine with what interest I read and studied it, and I must say that I think it is perfectly well done. The idea is like truth itself, so simple and obvious that those who read and understand it will be struck by its simplicity; and yet it is perfectly original. The reasoning is close and clear, and although so brief an essay, it is quite complete, embraces the whole difficulty, and anticipates and annihilates all objections. Few men will be in a condition to comprehend and appreciate the paper but it will infallibly create for you a high and sound reputation.[16]

Early in August 1857, one month after Bates's letter, the Dutch mail steamer delivered Darwin's first letter,[17] in which he informed Wallace that they had obviously thought much alike on the subject of species. Darwin's apparent insouciance was markedly different from Bates's recognition of the brilliance and originality of Wallace's ideas. Moreover, Darwin, despite including a coded warning to Wallace to keep off his patch, had asked for Wallace's help (Darwin wanted information about black jaguars' selection of mating partners, and about leopards on the islands he was exploring). Wallace probably wanted to think about how to reply to Darwin's letter, which remained unanswered for some time. There were things about it that bothered him.

The original letter, now part of the Addison Collection at the British Library, bears evidence of Wallace's reactions. He underlined Darwin's reference to his Sarawak Law, and his statement: 'I agree to the truth of almost every word of your paper'. He marked Darwin's comment that 1857 would be the '20th year' since he began work on his theory of species change. He underlined a sentence in which Darwin claimed that the body of his research had been 'how and in which way do species and varieties differ from each other'. Darwin wrote that a subject on which he had been experimenting and 'which cost me much trouble is the means of distribution of all organic beings found on ocean islands', and Wallace marked the passage with two strong parallel lines, obviously seeing this as a common area of interest.

However, Darwin's statement about the relationship between species and varieties must have particularly intrigued Wallace. If Wallace had commented on this matter in his first letter to Darwin, he would have outlined his belief that varieties and new species were one and the same. To find that Darwin's approach was to discover how they *differed* from each other must have concerned Wallace. Darwin had written that he and Wallace were thinking along much the same lines, so how could Darwin possibly believe that varieties and new species differed from each other?

In his reply to Bates, Wallace was keen that his friend should understand exactly how he now saw things. He wrote:

> To persons who have not thought much on the subject I fear my paper on the succession of species will not appear so clear as it does to you. That paper is, of course, only the announcement of the theory, not its development. I have prepared the plan and written portions of an extensive work embracing the subject in all its bearings... I have been much gratified by a letter from Darwin in which he says that he agrees with 'almost every word' of my paper. He is now preparing for publication his great work on

> species and varieties, for which he has been collecting information twenty years. He may save me the trouble of writing the second part of my hypothesis by proving that there is no difference in nature between the origin of species and varieties, or he may give me trouble by arriving at another conclusion, but at all events his facts will be given for me to work upon. Your collections and my own will furnish most valuable material to illustrate and prove the universal application of the hypothesis.[18]

Wallace would have known very well that his letter should have been with Darwin early in January 1857. The fact that it was delivered to Darwin four months late seems not to have troubled him. However, a glitch in Darwin's letter seems to reveal an attempt to explain why he had not replied earlier. Halfway through the first line, Darwin inserted the phrase 'received a few days ago' as an afterthought. It was a significant correction.

Shortly after receiving Darwin's letter, Wallace left for a new collecting site in the mountains thirty miles to the north of Macassar, but the collecting there was poor and in early September the rainy season set in. It seemed a good time to draft a reply to Darwin. He had much to report, given all his new insights on species and varieties and land bridges. The lack of response to his Sarawak Law still irked him, and along with his land bridge discovery he mentioned this disappointment to Darwin.

When the Dutch mail steamer left Macassar in October 1857, Wallace's second letter, dated 27 September, was on board. It was scheduled to arrive early in December, some nine weeks later. When Barbara Beddell went searching for this letter in Darwin's files in the mid-1960s, all that remained of it was one tiny section.[19] On one side of that clipped fragment of paper, Wallace had written that he had been most interested to learn from Darwin's letter that: 'my views on the order of succession of species were in accordance with your own for I had begun to be a little disappointed that my paper had neither excited discussion nor even elicited opposition.'

Darwin's deception had worked. Before Wallace's letter had arrived the previous January, Darwin had not understood that varieties and new species were the same thing,[20] nor did he have any evidential understanding of the importance of extinction, divergence and modification in nature to the evolutionary process, except for that already published by Wallace in his Sarawak Law and in the birds article.

On the other side of that carefully cut-out scrap of the letter was the information about black panthers that Darwin had requested. It was the only piece of the letter Darwin had kept, probably because it contained a valuable fact he might be able to use. Whatever else that it contained there is no absolute way of knowing. Just like the first letter almost a year before, there was nothing in it that Darwin wished to keep or anyone else to see.

CHAPTER 23

An insurance policy

DESPITE the shared intimacies of their most recent correspondence, in August 1857 Darwin and the American botanist Asa Gray were hardly confidantes. Darwin had been briefly introduced to Gray in 1839 by Joseph Hooker, and had met Gray and his wife again in 1851 at Hooker's house in Kew. Apart from these meetings, the men had spent no time together. However, since beginning his species work again, Darwin had picked Gray's brain on several occasions.

Now, though, Darwin's plan to tell Gray of 'his' new ideas was in its final stages. He must have known that he could not simply refer to the fact that varieties and new species were the same thing, or to his new-found realisation that organic change was not dependent on inorganic change, without citing some evidence. At the same time, he could not simply drop the theoretical ideas he had held onto for so long. He was still not convinced, despite Hooker's insistence, that his migration theory was flawed. He could add bits on to his old theory, but he was not certain enough of the concepts with which he was now dealing to make them the central planks of a new theory.

On 5 September 1857, he wrote one of the most important letters of his life.[1] He did what any pragmatist would do in his predicament. He first outlined aspects of his long-held theory, and then tagged his newly-discovered principle of divergence onto the end, almost as an afterthought. After one hundred and fifty years, it still reads oddly. Six months before, Darwin had written in the manuscript of 'Natural Selection' the words that dated his claim: 'The principle of divergence, I believe, plays an important part in the affinities or classification of all organic beings.' Yet he was still unable to fill in many details on which he based his entire theory. He suggested that Gray might like to do that himself.

In a personal note enclosed with the outline of his ideas, Darwin wrote:

> I forget the exact words which I used in my former letter, but I daresay I

> said that I thought you would utterly despise me, when I told you what views I had arrived at, which I did because I thought I was bound as an honest man to do so [...] I did not feel in the least sure that when you knew whither I was tending, that you might not think me so wild and foolish in my views (God knows arrived at slowly enough, and I hope conscientiously) that you would think me worth no more notice or assistance [...] I always expect my views to be received with contempt.
>
> As you seem interested in [the] subject, and as it is an *immense* advantage to me to write to you and to hear ever so briefly, what you think, I will enclose (*copied* so as to save you the trouble in reading) the briefest abstract of my notions on the means by which nature makes her species. Why I think that species have really changed depends on general facts in the affinities, embryology, rudimentary organs, geological history and geographical distribution of organic beings. In regard to my abstract you must take [it] immensely on trust; each paragraph occupying one or two chapters in my Book. You will, perhaps, think it paltry in me, when I ask you not to mention my doctrine; the reason is, if anyone, like the Author of the *Vestiges*, were to hear of them, he might easily work them in, and then I should have to quote from a work perhaps despised by naturalists and this would greatly injure any chance of my views being received by those alone whose opinion I value.

Darwin then sketched out his beliefs, which in essence are as follows:

1 That the organic form of domesticated animals can be changed by breeders selecting and accumulating variations in animals, caused by external conditions or genealogy.

2 That in nature, a change in the conditions of existence is the cause of a child not exactly resembling its parents. Geology suggests that such changes have been happening over millions of generations and are happening still.

3 That an unerring power called 'natural selection' has operated exclusively for the good of each organic being, but only a few of those born in each generation within a species survive to produce the next generation. There is some kind of check at work that is related to the struggle to survive.

4 That when environmental change occurs and the inhabitants struggle to survive, some individuals will be born with some slight variation. This variation, acted upon by natural selection, will allow the individuals to either co-exist with the parent form, or to survive by exterminating the parent form.

5 That any difficulties caused by this theory can be satisfactorily dealt with by saying that nature does not change its species by a series of jumps, but brings change about by a slow process over immeasurable time. The fact is that only some species are undergoing change at any one time. Any other objections can be put down to the imperfections of the geological record.
6 That a principle of divergence plays an important part in the origin of species. When this principle operates, each new variety or species will generally take the place of, and so exterminate, its less well-fitted parent.[2]

This, I believe, to be the origin of the classification or arrangement of all organic beings at all times. These always seem to branch and sub-branch like a tree from a common trunk; the flourishing twigs destroying the less vigorous, – the dead and lost branches rudely representing extinct genera and families.

This sketch is *most* imperfect; but in so short a space I cannot make it better. Your imagination must fill up many wide blanks. – Without some reflexion it will appear all rubbish; perhaps it will appear so after reflexion. – (C.D.)

He then tacked on a postscript:

This little abstract touches only on the accumulative power of natural selection, which I look at as by far the most important element in the production of new forms. The laws governing the incipient or primordial variation (unimportant except as to groundwork for selection to act on, in which respect it is all important) I shall discuss under several heads, but I can come, as you may well believe, only to very partial and imperfect conclusions.

This last confession must have alerted Gray to the idea that Darwin might have an idea of *what* happened, but was a long way from being able to reveal to him exactly *how* it happened.

Nowhere in the letter does Darwin explain how his principle of divergence works. A careful reading of his six points shows that he still considered the cumulative power of natural selection to be by far the most important element in the production of new species, rather than modification with descent away from an ancestral form. The principle of divergence was presented as something he had read but not yet understood. Certainly he had no example from nature of modification and divergence, unlike Wallace, who had used examples in his paper on birds published exactly one year earlier. For Darwin, that theoretical outline of 5 September must have been a very important document because he took the trouble of having a copy made of the letter and its six propositions. It was not Darwin's habit to copy his correspondence. During his lifetime, he wrote more than 14,000 letters but seems to have copied few others. Whatever his motivation at this time, nine months later it was to prove to have been a brilliant and inspired move as Darwin, shocked and desperate, sought help from his closest friends in an attempt to rescue his personal dream of scientific fame.

Gray would have received Darwin's letter by the end of September. His reply, which was with Darwin sometime before 29 November, could not have been very encouraging. What must have been evident to Gray was that Darwin's propositions simply were not convincing without supporting evidence. However, we will never know exactly what Gray said, because this was another of the letters Barbara Beddall went looking for in the mid-1960s and failed to find. This failure led her to state categorically that Darwin's letter to Gray was nothing more than an insurance policy to safeguard his claim to priority.

When Darwin replied to Gray on 29 November 1857, however, it was obvious that the reception his September letter had received was not the one he might have expected from a fellow doubter of the permanence of species. Yet it was one he might have expected from possibly the most respected botanist in America. Darwin thanked Gray:

> for your impression on my views. Every criticism from a good man is of value to me. What you hint at generally is very very true, that my work will be grievously hypothetical and large parts by no means worthy of being called inductive; my commonest error being probably induction from too few facts.[3]

He now offered Gray a definition of the principle of divergence: 'the tendency to the preservation from extinction of the most different members of each group'. This definition fell a long way short of Wallace's descent with modification away from an ancestral form.

Gray's response seems to have caused Darwin a period of doubt. He told his friend George Bentham at the beginning of December that he should expect to be disappointed by the book when it eventually appeared because, in an echo of Gray's criticism, it would be:

> grievously too hypothetical... It will very likely be of no other service than collocating some facts; though I myself think I see my way approximately

on the origin of species. But, alas, how frequent, how almost universal it is in an author to persuade himself of the truth of his own dogmas. My only hope is that I certainly see very many difficulties of gigantic stature. [4]

Nothing Darwin had offered between July and December 1857 seems to have made any great impression, so it would not have been the best time for him to receive Wallace's second letter.[5]

In all probability, this letter contained Wallace's convictions about his new proofs that varieties and species were the same thing, that modification and divergence now stood solidly at the heart of his theory, and that oceanic islands like Aru had been formerly connected to the mainland of New Guinea. The letter arrived at Down House on 6 December 1857, and Darwin's reply was dated 22 December.[6]

With his new ideas now indelibly on the record, 1858 might have looked encouraging for Darwin. However, his work on the 'Natural Selection' manuscript was slow, and had failed to impress his friends, especially Hooker. He had sent Hooker one of his chapters for help and criticism, and Hooker had chastised his method, telling Darwin that he had to be careful about his presentation of evidence and that he must draw his results from all kinds of evidence, both local and general, and use the knowledge of systematists from 'all shades of opinion'.[7]

On April 16, Darwin wrote to William Fox (his cousin, friend and collaborator) seeking help with the colour and markings of a particular breed of pony.[8] He was also having problems with the classification of large and small organisms, and was trying to finalise a section on bees. His letters give the impression of a man drowning in too much detail, still without a clear idea of where his research should take him, despite the clues he had already been given by Wallace (and almost everyone in England who was interested in aspects of the species question at the time). Following Hooker's advice about broadening his approach, he was also examining some bizarre possibilities. When he wrote to Hooker ten days after writing to Fox, he referred to the possibility of varieties being transferred from the Arctic to the Azores by icebergs;[9] his migration theory was still firmly in his thoughts.[10]

At the end of April 1858, continually troubled by ill-health, Darwin again left Down for a rest at Moor Park, a hydrotherapy centre in Farnham, thirty five miles to the southwest of London. Shortly after his arrival, he wrote to his wife: 'It was as pleasant a rural scene as ever I saw and I did not care one penny how any of the beasts or birds had been formed'.[11]

This statement does not shine with the confidence of a man who already knew the answer to that question, and the peace he enjoyed in the beautiful surroundings of Moor Park was the last he was to have for a long time. Charles Darwin was about to be forced to care, very deeply, about how beasts and birds were formed.

CHAPTER 24

Evolution solved

BACK IN the Malay Archipelago, Wallace had been thinking about food supplies, human existence and human species for almost a year. One day, he noticed islanders carrying their abundant burdens of rice and sago along the narrow path between two villages on the island of Gilolo, the most north-easterly of the Spice Islands.

When he had last written to Darwin, his question about whether Darwin would treat the subject of man in the book he was writing had not been idle. Wallace had become fascinated by man. Was every tribe in existence a different species? Were tribes varieties? Had extinction had the same effect on man as it had had on every other species?

He had come to Gilolo from the nearby island of Ternate to seek exciting new specimens and species, but it was the human tribes of Gilolo that absorbed Wallace's attention. He carefully noted everything he could about the people he had observed in the village of Dodinga: 'The village was entirely occupied by Ternate men, who are malays with an occasional admixture of papuan blood… The true indigenes of Gilolo [Papuans] live on the eastern coast or in the interior'.[1]

A few weeks later, he wrote:

> The natives of this large & almost unknown island were examined by me with much interest, as they would help to determine whether, independent of mixed races, there is any transition from the Malay to the papuan type. I was soon satisfied by the first half-dozen I saw that they were of genuine papuan race, lighter in colour indeed than usual but still presenting the marked characters of the type in features & stature. They are scarcely darker than dark Malays & even lighter than many of the coast malays who have some mixture of papuan blood. Neither is their hair frizzly or woolly, but merely crisp or waved, yet it has a roughness or slight woolliness, of appearance produced I think by the individual hairs not laying parallel & close together, which is very different from the smooth & glossy

> though coarse tresses, everywhere found in the unmixed malayan race. Their stature alone marks them as distinct being decidedly above the average malay height, while the features are as palpably unmalay as those of the European or the negro.[2]

Wallace was convinced he had finally found the dividing line between the two 'races': a place where the Papuan and Malay species of man were totally separated without any evidence of a transitional form existing between the two. Wallace believed that this fitted his theory. If all other animal species were defined by the extinction of intermediate forms, why not man himself? He had always considered man to be a species of animal, and thus subject to natural laws. He had seen how precarious was the existence of some of the tribes, especially with regard to food supplies, when he had visited Aru. Some of the scenes he had witnessed while living for some weeks with two separate tribes of Papuans on the northernmost island of the Aru chain had made him think deeply about humans' continuous quest for food.

In his field journal, he noted that he had lived with a tribe of Papuans in their landlocked village of Wokan for two weeks at the end of March 1857.[3] He compared that tribe's conditions of existence with those of another tribe of Papuans living in the coastal village of Wanumbai, with whom he lived for six weeks from the beginning of April.

Wallace made his feelings about man and beast clear from the outset. 'The human inhabitants of these forests are not less interesting to me than the feathered tribes,' he wrote in his journal, before addressing the condition of the Wokan villagers. 'They are on the whole a miserable set of savages. They live much as all people in the lowest state of human existence live and it seems to me now a more miserable life than ever I have thought it before'.

Part of their dilemma, he recognised, was that they had no staple food, such as bread, rice or sago. Instead, their diet consisted of various vegetables, plantains, yams, sweet potatoes and crude sago. They also chewed 'vast quantities' of sugar cane, as well as betel, gambir and tobacco. He noted that sometimes their diet was supplemented by cockles and other shellfish, or sporadic feasts of wild pig and kangaroo. These were 'too rarely to form anything like a regular part of their diet which is essentially vegetable, & what is of more importance as affecting their health, of green watery vegetables, roughly roasted or boiled & even this in varying & often insufficient quantities'.

Wallace thought the prevalence of skin disease and ulcers on the legs and joints of the villagers in Wokan were due to their inadequate diet. He had seen the same kind of thing among tribal people elsewhere, caused by the 'poorness and irregularity of their living'. Almost immediately, he realised that in the Amazon and now in the archipelago, tribes with staple diets were healthy and nourished, while those existing mostly on vegetables were in a poor condition.

> [I]n this as in other respects man does not seem capable of making a beast of himself with impunity living from hand to mouth on the herbs and fruits of the earth alone. He must labour and select and prepare some farinaceous product capable of being stored to give him a regular daily staple food. When this is obtained he may add vegetables and fruits with advantage, as well as animal food.

By contrast, in the village of Wanumbai, Wallace saw that the villagers were far removed from the 'miserable' existence of the people of Wokan:[4]

> They keep up a continual row from morning to night – talking laughing shouting without intermission; not very pleasant but I take it as a study of national character and submit… All the men and boys are expert archers never stirring out without their bows and arrows. They shoot all kinds of birds as well as pigs and kangaroos occasionally, which give them a pretty regular supply of meat with their vegetables. The result of this better living is superior healthiness, well made bodies and generally clear skin.

It was this energy and determination that Wallace recognised as the major difference between the two villages. His account makes his conclusions obvious: 'a people must exert diligence in procuring a steady food supply to maintain a state of healthiness.'

Wallace had also been disturbed by the way surface water disappeared through the 'porous coralline rock' of the island during any hint of a dry period. 'Were there a dry season like that of Macassar,' Wallace wrote, 'Aru would be uninhabitable'.[5] The inability of some tribes to organise for themselves the long-term provision of a staple foodstuff for their diet, and the vulnerability of the island-dwellers to anything like a long-term drought, dominated Wallace's thoughts after his first visit to Aru.

Therefore, one year later, in February 1858, observing Papuan villagers from the east of Gilolo delivering sago and rice in abundance to villagers in the west, the plight of some of those Aru islanders could not have been far from his mind, especially as he was still attempting to find evidence of a transitional race between the Papuans and the Malays.[6]

We have no contemporary time-frame for what happened next. There are no clues in Wallace's field journal, nor any in the retrospective accounts. However, during the time he spent on Gilolo during that February, something happened to Wallace's thinking that resulted in one of the most astonishing intellectual feats in the history of science.

Many years later, Wallace described how, while suffering from a 'sharp attack of intermittent fever which obliged me to lie down every afternoon during the cold and subsequent hot fits which lasted together two or three hours', the answer to the species question suddenly occurred to him.[7]

> It was during one of these fits, while I was thinking over the possible mode of origin of new species, that somehow my thought turned to the 'positive checks' to increase among savages and others described in much detail in the celebrated *Essay on Population* by Malthus, a work I had read a dozen years before.
>
> These checks – disease, famine, accidents, wars etc – are what keep down the population, and it suddenly occurred to me that in the case of wild animals these checks would act with much more severity, and as the lower order of animals all tended to increase more rapidly than man, while their population remained on the average constant, there suddenly flashed upon me the idea of the survival of the fittest – that these individuals which every year are removed by these causes – termed collectively 'the struggle for existence' – must on the average and in the long run be inferior in some one or more ways to those which managed to survive.
>
> The more I thought of this the more certain it appeared to be; while the only alternative theory – that those who succumbed to enemies, or want of food, or disease, drought or cold be every way and always as well constituted as those that survived – seemed to me impossible and unthinkable.

Wallace wrote that account in 1903, when he was a year away from his eightieth birthday, but the crucial elements of all five versions he wrote down between 1898 and 1908 are the same. In 1905, he commented on what was meant by the process he had described:[8]

> Then it suddenly flashed upon me that this self-acting process would necessarily improve the race, because in every generation the inferior would inevitably be killed off and the superior one would remain – that is, the fittest would survive. Then, at once, I seemed to see the whole effect of this, that when changes of land and sea, or of climate, or of food supply, or of enemies occurred – and we know that such changes have always been taking place – and considering the amount of individual variation that my

> experience as a collector has shown me to exist, then it followed that all the changes necessary for the adaptation of the species to the changing conditions would be brought about; and as great changes in the environment are always slow, there would be ample time for the change to be effected by the survival of the best fitted in every generation. In this way every part of an animal's organisation could be modified exactly as required and in the very process of this modification the unmodified would die out, and thus the definite characters and the clear isolation of each new species would be explained.

Wallace was convinced he was right:

> The more I thought it over the more I became convinced that I had at length found the long-sought-for law of nature that solved the problem of the origin of species. For the next hour I thought over the deficiencies in the theories of Lamarck and of the author of 'Vestiges' and I saw my new theory supplemented these views and obviated every important difficulty. I waited anxiously for the termination of my fit so that I might at once make notes for a paper on the subject. The same evening I did this pretty fully, and on the two succeeding evenings wrote it carefully in order to send it to Darwin by the next post which would leave in a day or two.

Wallace dated his paper 'Ternate, February 1858'.[9] It can be broken down into eight significant steps.[10]

1 Varieties can be expected to differ in organization and habits (and hence in their ability to gather food) from each other and from the 'parent species'.
2 Even slight differences will make a variety 'inferior' or 'superior' to the parent species under a given set of conditions.
3 The size of any population is a reflection of its food-gathering ability, not of its reproductive capacities.
4 Suppose, then, that a 'parent species' is represented in different geographical areas by an 'inferior' and a 'superior' variety. If the general circumstances worsen, as in a prolonged drought that makes food scarce, the populations of all three will dwindle.
5 However, the variety that is inferior in its food-gathering abilities (and consequently has a smaller population) will be the first to dwindle to extinction, followed by the parent species. At this point, only the superior variety will survive.
6 If conditions improve, and food becomes abundant, the population of

> the surviving variety will increase and extend its range, eventually attaining the size and range of the three former populations.
> 7 Thus the superior variety replaces the parent species, becoming what must be called a 'new' species, and in time becoming a 'parent', geographical representatives of which may become new varieties.
> 8 The repetition of the process results in progressive development and continued divergence from the original type.

Wallace's paper was entitled 'On the Tendency of Varieties to Depart Indefinitely from the Original Type'. It was nothing less than a perfect description of the process of evolution, completed in just over four thousand words (See Appendix 4).

The Dutch mail steamer *Ambon*,[11] with the letter on board, departed from Ternate on 9 March 1858, on the first leg of a journey to England and the home of the unsuspecting Charles Darwin.

CHAPTER 25

Animal trails

JOHN BROOKS, the Yale environmental biologist who had stumbled on similarities between the ideas of Darwin and Wallace while preparing his lecture notes, had long been aware that there was a previously unidentified anomaly in Darwin's 'Natural Selection' manuscript.[1] At the end of March 1857, Darwin finished writing the 'Natural selection' chapter itself, which included the very brief two pages in which he committed himself to the principle of divergence.

He also knew that Darwin had gone back to amend this chapter between May and June the following year, and in doing so had deleted these two pages and replaced them with an entirely new section of 41 pages. Brooks was intrigued. What new information did Darwin have access to in the early summer of 1858 that would have caused him to return to a chapter he had completed more than a year before?

Brooks knew that in his letter to Gray the previous September, Darwin had referred to the principle of divergence, which he considered so important for his 'notions on the means by which nature makes her species', but he had not spelled out exactly how divergence worked. How important had Darwin considered this principle to be when he had first written about it? If Darwin's ideas had changed so much since he finished the 'natural selection' chapter in March 1857, why had he not added this 41-page section to his manuscript when he wrote to Gray that September, rather than waiting until the following summer? It did not make sense to Brooks, but he was convinced about one thing. He was sure that he would find that the 41-page section dealt with divergence.[2] By studying a microfilm copy of Darwin's 'Natural Selection' manuscript, Brooks quickly established that his suspicion was correct.

A cursory examination of the microfilm convinced Brooks that the contents of the inserted divergence material were similar to the last twenty pages of Chapter IV of *Origin*. A closer examination revealed that those pages had unquestionably been based on the 41-page insertion.[3]

However, there was something else, which Brooks had not expected. In the same chapter of the manuscript, only a few pages further on, there was a second insertion. Darwin had added a further 25 pages under the heading: 'On the absence of intermediate forms or Links between species of the same Genus'. After writing this new section and inserting it into his manuscript, Darwin had added the following footnote: 'June 1858. I doubt whether I have got intermediate links quite clear'.[4] So when had Darwin written out these two large passages of additional material and do they amount to evidence of Darwin's plagiarism of Wallace's work?

Inspecting the original 'Natural Selection' manuscript at Cambridge sometime later, Brooks was amazed to find how much the inserted material stood out. All the original pages of the manuscript had been written on a grey wove paper, but the two long interpolations Darwin had added in June 1858 had been written on a bluish grey paper with only two grey wove original pages surviving. The additions were obvious. The forensic case had been made.[5]

By the absence of a whole page from the original manuscript and some crossings out on the two adjacent pages, Brooks estimated that the original length of the section headed 'Extinction' had been about the same length as the section in Darwin's letter to Asa Gray. Even the words of one brief section that remained were similar. In the original text of his Natural Selection sketch, Darwin had written: 'The principle of divergence, I believe, plays an important part in the classification of all living beings'. Six months later, in his letter to Gray, he had written, 'This I believe to be the origin of the classification and affinities of organic beings at all times.'

Brooks believed that it was reasonable to assume that the rest of the text excised from the manuscript had been similar in content to that of the Gray letter. He was also convinced that Darwin's treatment of the subjects of extinction, intermediates and the natural system of affinities in the later chapters of the *Origin of Species* could in large part be matched with the material that Darwin had added to his manuscript less than a fortnight before 18 June, the date – he told Lyell – on which he had received Wallace's third letter.[6]

For Brooks, both sections indicated that his suspicions were justified, and that the similarities he had seen between the written accounts of Wallace and Darwin were not imaginary. But he was still puzzled: the 41-page addition on divergence bore similarities not just to the Sarawak Law, but to something else.[7]

Brooks was wondering why Darwin had gone back to re-read the Sarawak Law paper before making those insertions, when a new thought crossed his mind. In the Sarawak Law paper, Wallace had speculated about the roles of extinction, modification and divergence in creating the natural patterns of affinity. The most likely stimulus to have caused Darwin to think again about those ideas would have been Wallace's latest theory, written in February 1858 and mailed from Ternate on 9 March. In that manuscript, Wallace had presented simple natural dynamics that accounted for extinction, modification, divergence and the origin of new species.

But how could Wallace's Ternate paper have influenced Darwin's thinking when his letter did not arrive at Darwin's house until 18 June, some time after Darwin had written the new material into his manuscript?[8]

The implication could not be ignored. What if this letter had arrived at Darwin's home earlier than the date he claimed?

What if Darwin had again used Wallace's ideas, this time to complete the divergence section of his 'Natural Selection' manuscript, and hidden that fact from both Lyell and Hooker?

Brooks knew that he could not tackle this question from his study at Yale. Answers, if there were any, were to be found only in obscure shipping, postal and newspaper archives at least a hundred and fifty years old, and three thousand miles away. He also knew that asking such a question was not going to make him popular, but he could not simply close a line of research that promised to change the scientific record. He visited the Post Office archives in Mount Pleasant, London, and the Dutch mail archives at the Postmuseum in The Hague, and slowly pieced together the most obscure elements of the British Post Office's delivery schedule for letters posted to and from the Malay Archipelago during the relevant period. He also searched through the archives of the P&O shipping line, which had been commissioned by the Admiralty on behalf of the British and Dutch governments to carry the mail to and from the Far East.

In one summer's research, he dredged up detailed evidence not only of exactly where letters had been at every stage of the journey from Singapore to London, but also of the Dutch mail steamer service's regular monthly journeys around the islands of the archipelago. By the time he had finished his investigations, Brooks had a unique contemporary record of the journeys made between Singapore and London by P&O ships for the crucial months from March to June 1858.[9] He had done the groundwork for building a comprehensive case that Darwin had lied about when he received Wallace's third and most important letter. Unfortunately for Brooks, he did not pay close enough attention to one of the records he had uncovered, and in the process introduced an inconsistency between his conclusions and those of McKinney. Darwin's supporters latched on to this inconsistency and used it to try to diminish the validity of all his work. Instead of checking the records themselves, Darwin's followers believed that they had seen off another attack on their hero.

However, the path that John Brooks signposted in his search for what really happened to Wallace's letter was not a dead end. On the contrary, if other Darwin researchers had followed the leads that Brooks had uncovered they would have been surprised to find there existed overwhelming evidence that Wallace's third letter from Ternate on 9 March 1858 had also arrived weeks earlier than Darwin claimed.

CHAPTER 26

Wallace's first letter

SUBSTANTIATING the charge that Charles Darwin stole crucial material from Wallace and claimed it as his own relies on the evidence of Darwin's own collected correspondence and insertions into his 'Natural Selection' manuscript, which record how his understanding of the species question changed between January 1857 and July 1858.

Most crucially, it relies on the five letters exchanged between the two men, and the fact that Darwin had in his possession letters from Wallace that he claimed arrived much later than they should have done.

It all started with the first unsolicited letter Wallace had written to Darwin. On the basis of the documented shipping schedules of the P&O mail delivery and collection service in operation at the time, and archival records (held at the Post Office museum in London) of when the post was delivered to addressees, it is possible to say with confidence that the letter Wallace wrote to Darwin from Macassar (dated 10 October 1856) took the following route on these dates:[1]

Timeline of the first letter to Darwin

- The letter was aboard the *Makassar*, which left Macassar on 31 October. This ship arrived in Surabaya, Java, on 2 November.
- Mail from the *Makassar* was transferred to the *Koningin der Nederlanden*, which left for Batavia (now Jakarta) on 4 November.
- The *Koningin der Nederlanden* arrived in Batavia on 7 November and departed for Singapore on 11 November.
- The *Koningin der Nederlanden* arrived in Singapore on 15 November.
- Mail was transferred to the P&O steamer *Singapore*, which left Singapore on 23 November and arrived in Galle, on the southern coast of Ceylon, on 1 December.

- The mail was then transferred to the *Bengal*, which left Galle for Suez on 1 December.
- The *Bengal* arrived in Suez on 15 December.
- The mail was transferred by train to Alexandria and loaded on board the steamer *Ripon*.
- The *Ripon* left Alexandria on 26 December and docked briefly in Malta on 30 December.
- Heavy storms had delayed the *Ripon*, which was due to arrive in Southampton on 2 January. It finally arrived on 11 January.
- Wallace's letter to Darwin would have been delivered on 12 January 1857.

Darwin claimed that this letter arrived in late April 1857.

CHAPTER 27

A second letter

WALLACE wrote his second letter to Darwin on 27 September 1857.[1] The sole remaining fragment of that letter records Wallace's unhappiness that no one had responded to his Sarawak Law. In his reply, sometime after receiving the letter on 6 December that year, Darwin assured Wallace that his Sarawak Law paper had been commended to him by Sir Charles Lyell and Edward Blyth. He then wrote something much more revealing. Wallace had obviously told him of his belief that the island of Aru had once been joined to New Guinea, until the land between had dropped beneath the ocean, leaving animals of the same species common to both islands. Darwin told Wallace at the end of the letter that he would fight to the death against this explanation of how species got to oceanic islands.

This suggests that for a long time – certainly up to the end of 1857, and probably up until Wallace's Ternate paper arrived at his home – Darwin was still convinced that his migration theory of species was a better explanation of the geographical distribution of animals and plants than anything either Forbes or Wallace was proposing.

His dismissal of Forbes's continental extension idea, when Wallace was offering him a powerful proposition that echoed Forbes's botanical evidence and Wollaston's land-mollusc evidence, suggests a reluctance in Darwin to change ideas. It seems that up until the middle of 1858, Darwin was still unprepared to consider stronger and more persuasive alternative explanations of geographical distribution, despite having told Asa Gray that he considered his principle of divergence responsible for the creation of new species.

Darwin, in tacking divergence onto more traditional thinking, was claiming the idea as an insurance without having a great deal of confidence in its power or a proper understanding of how it was being used by Wallace. Certainly, Barbara Beddall was not convinced that Darwin really understood divergence by the time *On the Origin of Species* was due to be published.[2]

Wallace's second letter had arrived at Darwin's home on 6 December 1857. Darwin had written and posted his reply by 22 December, and sent it first class. The P&O steamer *Pera* with the second class mail had already left Southampton for Malta on 20 December[3]. Darwin's letter was carried by a combination of boats and trains via Marseilles to Malta, where it was taken on board. The boat sailed immediately for Alexandria. The mail was taken overland to Suez on the Red Sea, and placed on the P&O liner for the journey to Singapore.

On 1 February 1858 in Singapore, Darwin's letter to Wallace was transferred to the waiting Dutch steamer *Banda*,[4] which left the next day for Batavia, the largest town on the island of Java, where it arrived on 7 February. The following day, the mail was transported by the *Banda* to the port of Surabaya, five steaming days along the coast to the east of Batavia, where it was put aboard the *Ambon*, the inter-island mail steamer. The next day, the *Ambon* left for Macassar on the island of Celebes. It was 14 February 1858.[5] It arrived at Macassar a few days later, and then on 9 March it reached Ternate, where Alfred Wallace was waiting at the harbour to take delivery of some boxes. The boxes[6] did not arrive on that boat, but he was also waiting to place on board his letter to Frederick Bates in Leicester, and his third letter to Darwin containing his Ternate Law paper.

In his journal he made no mention of the unexpected arrival of Darwin's letter, but he must have opened it and read it at the quayside because of the reference to Lyell. Darwin had written:

> My dear Sir,
> I thank you for your letter of Sept 27th. – I am extremely glad to hear that you are attending to distribution in accordance with theoretical ideas. I am a firm believer, that without speculation there is no good and original observation. Few travellers have [at]tended to such points as you are now at work on; and indeed the whole subject of distribution of animals is dreadfully behind that of Plants. – You say that you have been somewhat surprised at no notice having been taken of your paper in the *Annals*: I cannot say that I am; for so very few naturalists care for anything beyond the mere description of species. But you must not suppose that your paper has not been attended to: two very good men, Sir C. Lyell, and Mr E Blyth at Calcutta, specially called my attention to it. Though agreeing with you on your conclusion[s] in that paper, I believe I go much further than you; but it is too long a subject to enter on my speculative notions.
>
> I have not yet seen your paper on distribution of animals in the Arru Islands. – I shall read it with *utmost* interest … but I can see that you are

inclined to go much further than I am in regard to the former connections of oceanic islands with continent: Ever since poor E. Forbes propounded this doctrine, it has been eagerly followed; and Hooker elaborately discusses the former connections of all the Antarctic islands and New Zealand and S. America – About a year ago I discussed this subject much with Lyell and Hooker (for I shall have to treat of it) and wrote out my arguments in opposition: but you will be glad to hear that neither Lyell nor Hook[er] thought much of my arguments: nevertheless for once in my life I dare withstand the almost preternatural sagacity of Lyell. –

You ask about land-shells on islands far distant from continents: Madeira has a few identical with those of Europe, and here the evidence is really good as some of them are sub-fossil. In the Pacific islands there are cases, of identity, which I cannot at present persuade myself to account for by introduction through man's agency; although Dr. Aug. Gould has conclusively shown that many land-shells have there been distributed over the Pacific by man's agency. These cases of introduction are most plaguing. Have you not found it so, in the Malay Archipelago? It has seemed to me in the lists of mammals of Timor and other islands, that *several* in all probability have been naturalised. –

Since writing before, I have experimentised a little on some land-mollusca and have found sea-water not quite as deadly as I anticipated. You ask whether I shall discuss 'man'. – I think I shall avoid [the] whole subject, as so surrounded with prejudices, though I fully admit that it is the highest and most interesting problem for the naturalist. – My work, on which I have now been at work more or less for 20 years, will *not* fix or settle anything; but I hope it will aid by giving a large collection of facts with one definite end: I get on very slowly, partly from ill-health, partly from being a very slow worker. – I have got about half written; but I do not suppose I shall publish under a couple of years. I have now been three whole months on one chapter on Hybridism! –

I am astonished to see that you expect to remain out 3 or 4 years more; what a wonderful deal you will have seen; and what interesting areas, - the grand Malay Archipelago and the richest parts of S. America! – I infinitely admire and honour your zeal and courage in the good cause of Natural Science; and you have my very sincere and cordial good wishes for success of all kinds; and may all your theories succeed, except that on oceanic islands, on which subject I will do battle to the death.

Pray believe me, My dear Sir, Yours very sincerely, C. Darwin

Wallace had nothing to lose. Charles Lyell had figured so much in his ideas and also in his objections over the years that he must have been astonished to find that Lyell had taken the trouble to commend his Sarawak Law paper to Darwin. His covering letter to Darwin was already written, but if Lyell was interested in his ideas, why not suggest that Darwin, if he thought well of his paper, forward it to Charles Lyell for his opinion? It would have been an on-the-spot decision by Wallace, because he knew neither man.[7] His only contacts with Darwin had been the two letters they had exchanged, and he could have had no idea that Lyell had thought well of his Sarawak Law until the moment he read Darwin's letter.

When the steamer left Ternate, Wallace's third letter to Darwin, together with a letter to Henry Bates via his brother in Leicester, was bundled with the rest of the European mail, heading for England under lock and key.

CHAPTER 28

The third letter

WALLACE'S advocate H. L. McKinney completed his research, gave his lectures and wrote his book, but he was never going to win prizes for generosity in sharing his sources. In the late 1970s, Arnold C. Brackman, a US writer and journalist who specialised in historical narrative, was researching a book on Wallace. He contacted McKinney congratulating him on his own book, and asking for some help.[1]

Brackman wanted to know how he could get in contact with Wallace's descendents (who had helped McKinney some years earlier) so that he could check for himself an incredible piece of evidence McKinney had been shown. McKinney did not reply. Brackman wrote again. Again McKinney failed to reply. Brackman was furious, and used the story in his book in an attempt to embarrass McKinney within the academic fraternity and with the government agencies that had funded his research. McKinney still did not get in touch. It was obvious that as far as he was concerned, the Wallace story was his personal preserve. If Brackman so badly wanted to check his evidence, he would have to find the Wallace family himself.[2]

Brackman spent hours with the London telephone directory and failed with every 'Wallace' he contacted. Then he stumbled upon a reference to the Wallace family's connection with Bournemouth, a coastal resort in the south of England, and switched his point of attack. In a very short space of time, he was sitting with the family in their home. The object that he had come to see was old and fragile, but he immediately saw why McKinney had made it central to the writing of his book.

The discovery was a letter from Wallace addressed to Frederick Bates, the brother of Henry Bates, at his home in the English town of Leicester.[3] The letter, dated 2 March 1858, had originally enclosed Wallace's January letter to Henry Bates in the Amazon (in which he had informed his friend that he had received a letter from Darwin), which had been forwarded to Bates as Wallace had requested. The envelope was marked 'via Southampton' in Wallace's handwriting, and bore a cancellation mark from Singapore dated 21 April. It also bore a London postmark with the date 3 June 1858 (as well as one for the same date at Leicester).

McKinney had immediately recognised the significance of these dates when he had first seen the envelope years before. Darwin had written to Hooker on 8 June 1858, saying that the principle of divergence, along with natural selection, was now the keystone of his theory. Four days after that, Darwin had written in his journal that he had finished amending the 'Natural selection' chapter.

Darwin had claimed that Wallace's letter had arrived at his home on 18 June, but the amazing piece of evidence Brackman was holding suggested otherwise. If this letter to Frederick Bates had arrived in London by 3 June, then so, too, had Wallace's letter to Charles Darwin. They had travelled on the same ship from Singapore, under a Royal Navy guard.

For Brackman, this piece of evidence (along with Beddall's suspicions about the missing letters to and from Darwin in that crucial three-year period) indicated a conspiracy he had never dreamed of when he had begun his biography of Wallace. When Brackman published his book, which gave details of evidence that pointed to Darwin having lied about when he had received Wallace's Ternate paper, the Yale environmental biologist John Brooks was still in the middle of his research, and could tell Brackman only that he, too, had a strong suspicion that Wallace's letter had been in Darwin's hands weeks earlier than the declared arrival date.

The question could not be avoided then and it cannot be avoided now: what evidence is there that Darwin stole Wallace's ideas from his Ternate paper and claimed them as his own by incorporating them in his 'Natural Selection' manuscript between 3 June and 18 June 1858?

By piecing together nineteenth century shipping reports from the Dutch East Indies and re-examining Post Office archives in London, as well as following leads from shipping experts for the period, it is now possible to claim beyond any reasonable doubt the following course of events:[4]

Timeline of the third letter to Darwin

- Wallace posted letters to both Charles Darwin and Frederick Bates on 9 March 1858 from Ternate.
- Both letters were carried aboard the *Ambon* and arrived in Macassar on 20 March.

- The *Ambon* carried the letters from the Celebes to the Javan port of Surabaya and arrived on 25 March.
- The letters were transferred to the *Koningin der Nederlanden*, which arrived in Batavia on 31 March.
- The *Koningin der Nederlanden* sailed from Batavia on 12 April and arrived in Singapore on 16 April.
- On 21 April, both letters were date-stamped in Singapore and put on board the *Bombay* en route to Ceylon. The *Bombay* arrived in Galle on 29 April.
- At Galle, the letters were transferred to the *Nubia*, which set sail for Suez on 2 May.
- The *Nubia* arrived in Suez on 16 May.
- In Suez, the letters were transferred by the overland route and arrived in Alexandria on 18 May.
- The *Pera* picked up the mail at Alexandria and sailed on 19 May for Malta. The ship docked for three hours in Malta on 23 May and set sail for Southampton.
- The *Pera* arrived in Southampton, on schedule, shortly after noon on 2 June.
- Mail from the *Pera* was despatched from Southampton to London by 3pm on 2 June.
- The *Pera* mail arrived in the General Post Office in London at 6.30pm on 2 June.
- Wallace's letters were date-stamped and delivered to Charles Darwin and Frederick Bates on 3 June.

The arrival of Wallace's letter on 3 June (see Appendix 2) would have given Darwin more than enough time to digest its contents and make the two lengthy changes to the 'natural selection' chapter of his manuscript. It would also have allowed him to claim that Wallace's ideas were replicas of his own.

This explains why Darwin had suddenly been able to expound on his ideas about divergence (increasing the relevant section of the manuscript from fewer than two pages to more than sixty pages in the first two weeks of June 1858), and how he had got hold of the essence of Wallace's ideas about 'intermediate types and links of species', which Wallace had been thinking and writing about just before he sent off his letter to Darwin.

It was all extremely neat, and though John Brooks's date differed from McKinney's, he believed he had enough evidence to accuse Charles Darwin of stealing Wallace's ideas in a way that McKinney had not, because McKinney had missed the implications of the changes Darwin had made to his manuscript in June 1858.

Only now, when the entire journey of those letters can be verified beyond doubt, can the case be made that ideas contained in Wallace's Ternate Law paper were plagiarised by Charles Darwin to convince both Lyell and Hooker, should such evidence ever have been needed, that he had already described those ideas in his own 'Natural Selection' manuscript some time before Wallace's letter arrived at his home.

CHAPTER 29

Checkmate

SIR CHARLES LYELL and Dr. Joseph Hooker were the first members to enter the meeting room at the Linnean Society in central London on the afternoon of 1 July 1858. Both men were registered present. Lyell's name headed the list. It was the first time in five years he had attended a meeting of the society. Hooker's name followed immediately; it was his first meeting for a year.[1]

Both men had to have been aware of the questionable nature of the service they were about to undertake. As their colleagues arrived for the meeting and the room began to fill, Lyell in particular must have wondered how Darwin had managed to get himself into such a situation.

Their relationship over recent years had been nothing like as close as it had once been. Two years before, he had warned Darwin that Wallace was running him close in the race to solve the mystery of the origin of species. Since then, he, unlike Hooker, had not been kept up to date with Darwin's ideas, although he had attempted to straighten out his protégé's thinking as Darwin wrestled with certain ideas, such as Forbes's Atlantis theory.

Lyell would have been surprised when he received Darwin's letter enclosing Wallace's Ternate Law. Wallace had asked Darwin, assuming that the latter thought Wallace's theory sufficiently novel and interesting, to send it on to Lyell. Lyell knew that many of Wallace's original ideas, in both the Sarawak and Ternate papers, had been at least partly prompted by the minute study of his own *Principles*, but until Darwin told him that he had mentioned his approval of the Sarawak paper to Wallace, Lyell had not known that Darwin and Wallace were corresponding.

He must have remembered clearly Darwin's devastation in his first letter after receiving Wallace's Ternate paper two weeks before:[2]

> My dear Lyell,
> Some year or so ago, you recommended me to read a paper by Wallace in

the *Annals*, which had interested you & as I was writing to him, I knew this would please him much, so I told him. He has today sent me the enclosed & asked me to forward it to you. It seems to me well worth reading. Your words have come true with a vengeance that I should be forestalled. You said this when I explained to you here very briefly my views of 'Natural Selection' depending on the Struggle for existence. – I never saw a more striking coincidence. If Wallace had my M.S. sketch written out in 1842 he could not have made a better short abstract! Even his terms now stand as Heads of my Chapters.

Please return me the M.S. which he does not say he wishes me to publish; but I shall of course at once write & offer to send to any Journal. So all my originality, whatever it may amount to, will be smashed. Though my Book, if it will ever have any value, will not be deteriorated; as all the labour consists in the application of the theory.

I hope you will approve of Wallace's sketch, that I may tell him what you say.

My dear Lyell,
Yours most truly,
C.Darwin

How Lyell reacted when he read this letter, we will probably never know. Certainly, he would have had cause for some kind of remonstration with Darwin, written or verbal, but Lyell's reply no longer exists and there is no record of any meeting between the two men during this period. One week later, Darwin wrote his second letter to Lyell:[3]

My dear Lyell,
I am very very sorry to trouble you, busy as you are, in so merely personal an affair. But if you will give me your deliberate opinion you will do me as great a service, as ever man did, for I have entire confidence in your judgment & honour…

There is nothing in Wallace's sketch which is not written out much fuller in my sketch copied in 1844, & read by Hooker some dozen years ago. About a year ago I sent a short sketch of which I have [a] copy of my views (owing to correspondence on several points) to Asa Gray, so that I could most truly say & prove that I take nothing from Wallace. I should be *extremely* glad <u>now</u> to publish a sketch of my general views in about a dozen pages or so. But I cannot persuade myself that I can do so honourably. Wallace says nothing about publication, & I enclose his letter. But as I had not intended to publish my sketch, can I do so honourably because Wallace has sent me an outline of his doctrine? – I would far rather burn my whole book than that he or any man should think that I had behaved in a paltry spirit. Do you not think his having sent me this sketch ties my hands? I do not in least believe that he originated his views from anything which I wrote to him.

If I could honourably publish I would state that I was induced now to publish a sketch (and I should be very glad to be permitted to say to follow your advice long ago given) from Wallace having sent me an outline of my general conclusions. – We differ only, that I was led to my views from what artificial selection has done for domestic animals. I could send Wallace a copy of my letter to Asa Gray to show him that I had not stolen his doctrine. But I cannot tell whether to publish now would not be base & paltry: this was my first impression, & I should have certainly acted on it, had it not been for your letter.

This is a trumpery affair to trouble you with; but you cannot tell how much obliged I should be for your advice. –

By the way would you object to send this & your answer to Hooker to be forwarded to me, for then I shall have the opinion of my two best and kindest friends. – This letter is miserably written & I write it now, that I may for [a] time banish [the] whole subject. And I am worn out with musing.

I fear we have a case of scarlet-fever in [the] house with Baby. Etty is weak but is recovering. –

My good dear friend forgive me. – This is a trumpery letter influenced by trumpery feelings.

Yours most truly,

C. Darwin

I will never trouble you or Hooker on this subject again. –

The next day, Darwin wrote again to Lyell:[4]

My dear Lyell

Forgive me for adding P.S. to make the case as strong as possible against myself.

Wallace might say 'you did not intend publishing an abstract of your views till you received my communication, is it fair to take advantage of my having freely, though unasked, communicated to you my ideas, & thus prevent me forestalling you?' The advantage which I should take being that I am induced to publish from privately knowing that Wallace is in the field. It seems hard on me that I should be thus compelled to lose my priority of many years standing, but I cannot feel at all sure that this alters the justice of the case. First impressions are generally right & I at first thought it would be dishonourable in me now to publish. –

Yours most truly C. Darwin

I have always thought you would have made a first-rate Lord Chancellor; & I now appeal to you as a Lord Chancellor.

It seems that it was at this stage that Lyell decided to act to protect Darwin's interests. His letter in reply was sent via Kew for Hooker to read. It seems that Hooker then added a letter of his own, because on 29 June, mourning the death of his youngest son, Darwin responded to a request from Hooker for a copy of the letter to Asa Gray.[5] The letters from both Lyell and Hooker are missing.

My dearest Hooker,

I have received your letters. I cannot think now on [the] subject, but soon will. But I can see that you have acted with more kindness & so has Lyell even than I could have expected from you both most kind as you are.

I can easily get my letter to Asa Gray copied, but it is too short. –

Poor Emma behaved nobly & how she stood it all I cannot conceive. It was wonderful relief when she could let her feelings break forth. –

God bless you. – You shall hear soon, as soon as I can think.

Yours affectionately,

C.Darwin.

A further letter the same night to Hooker suggests that his friend had written back immediately, but this letter is also missing:[6]

My dear Hooker,

I have just read your letter, & see you want papers at once. I am quite prostrated & can do nothing but I send Wallace & my abstract of abstract of letter to Asa Gray; which gives most imperfectly <u>only</u> the *means of change & does not touch* on [the] reasons for believing species do change. I daresay all is too late. I hardly care about it. –

But you are too generous to sacrifice so much time & kindness. – It is most generous, most kind. I send sketch of 1844 <u>solely</u> that you may see by your own handwriting that you did read it. –

I really cannot bear to look at it. – Do not waste much time. It is miserable in me to care at all about priority. –

The table of contents will show what it is. I would make a similar but shorter & more accurate sketch for [the] Linnean Journal. – I will do anything.

God bless you my dear kind friend. I can write no more. I send this by servant to Kew.

Yours, C. Darwin.

And so, on that first day of July, after a hectic two weeks in which extracts had been copied, evidence collated and arguments prepared, Lyell and Hooker were ready to address the special meeting of the Linnean Society at its premises off Piccadilly in central London.

Neither the grieving Darwin nor Wallace, the working collector, was present. That was to the advantage of Darwin's two closest friends. Any statements they made would not be challenged, and every statement had only one aim: to ensure priority for Darwin, and by so doing ensure his permanent place in scientific history.[7] They invited their fellow members to accept that he had long espoused certain ideas and doctrines, thus establishing his claim to priority.

To challenge Wallace's excellent and comprehensive paper, they used three pieces of evidence selected from Darwin's archive. The first two were brief extracts from the unpublished essay of 1844. The third was the letter to Asa Gray, written only nine months before. No extracts from the manuscript of 'Natural Selection' were used, despite the fact that it contained Darwin's most recent ideas about divergence and varieties, added to the manuscript only weeks before.

Lyell and Hooker had decided that they would put their own reputations on the line. They agreed, as Darwin was now claiming, that he had sketched out his evolutionary theory not in 1842, but in 1839. Moreover, they claimed that the contents of the 1844 essay had not only been read by Hooker, but had also been communicated to Lyell himself. There was no mention of the fact that both men had voiced serious objections over several years to Darwin's migration theory, which (alongside the idea of perfect adaptation) had been central to his thinking in 1844 and for a long time afterwards.

Then, they read the extracts from the 1844 essay. The first was from a section dealing with variation in nature, highlighting Darwin's concept of natural selection and a comparison between domestic races and true species. The second, perhaps the most apposite section they could have selected, established that Darwin had been familiar with the ideas of Thomas Malthus long before he had written the essay. Lyell and Hooker had evidently planned this to echo Wallace's claim in his own paper that the ideas of Malthus had led him directly to his complete understanding of the process of evolution. In fact, of course, the previous year Darwin had indicated that the greatest influences on his ideas had been the Reverend W. Herbert, de Candolle and Lyell himself. Malthus's name had not even been mentioned in his letter to Gray.

Lyell and Hooker then tried to establish that in September the previous year, Darwin had written the letter to Gray indicating his awareness of what he called 'a principle of divergence', which he believed to be the origin of the classification and arrangement of all organic beings at all times. The Gray letter was the crucial piece of evidence, and the two men hoped it would serve to convince their fellow Linnean members that Darwin had solved the problem of how species evolve long before he received the crucial letter from Wallace.

With this selection of evidence placed firmly on the record before Wallace's paper was read out to the fellows, Lyell, who had trained as a lawyer, hoped to convince his audience that it was Darwin who had led, and Wallace who had followed.

Despite their strategy, Lyell and Hooker must have expected some awkward questions. They would have been aware that some might argue that a complete scientific paper should always be read before a selection of extracts. At the meeting, however, no objection to procedure was offered.

There were extenuating circumstances. This was not a normal meeting, in which scientific knowledge was disseminated and discussed. The society's scheduled meetings for the year had already been completed. This particular meeting had been called to discuss the election of a new Vice-President to replace Sir Robert Brown, who had died. The Darwin/Wallace priority discussion had been a last-minute addition to the business planned for that day.

After Lyell and Hooker had finished their presentation, there was absolutely no discussion. The members agreed that Darwin and Wallace should be acknowledged as co-discoverers of the theory of how species evolve, which would henceforth be known as the Darwin–Wallace theory of evolution. The crucial question of priority was settled by placing Darwin's name before Wallace's. Lyell and Hooker had successfully conspired to hand Charles Darwin the prize he had coveted for more than twenty years.

Following the meeting, Lyell returned to his home in central London, Hooker renewed his scientific and administrative work at Kew Gardens, and Charles Darwin urgently began writing an abstract of his 'Natural Selection' manuscript for his publisher John Murray. The following month, Wallace's theory of evolution was published in the Linnean Society's journal.[8] *On the Origin of Species* was not published until more than a year later.

Three months after Lyell and Hooker had secured priority for Darwin, Wallace learned of the events at the Linnean Society from a letter written by Hooker. Darwin had felt incapable of doing the job himself, and had asked Hooker to write to Wallace to inform him of the meeting and its outcome. Hooker had never previously communicated with Wallace.[9]

He need not have worried. Wallace, unassuming, lonely and ignored for so long, was absolutely delighted to have his name conjoined with that of Darwin. He never found out exactly what had happened in the early summer of 1858 after his letter arrived at the home of the great Charles Darwin, or how the ideas he expressed in his letter of October 1856 had given Darwin insights that had eluded him since 1837.

In November 1859, little more than a year later, Darwin published *On the Origin of Species*.[10] In the body of the text, the name of Alfred Russel Wallace appeared three times, but only in the context of his diligence as a researcher and observer. Nowhere in the book was he acknowledged as the co-discoverer of the solution to the mystery of evolution, nor was there any mention of his brilliant Ternate paper.

For that, Wallace would have to wait.

CHAPTER 30

End of a conspiracy

FOUR DAYS after the Linnean meeting, and assured of his priority, Charles Darwin wrote another letter to Asa Gray in Boston.[1] Darwin asked about the precise date of the letter he had sent to Gray ten months before, in which he had spelled out his ideas on species. He refers to 'a letter written in September, October or November'. This is strange, because Darwin knew exactly when he had sent it; he had kept the copy that was to prove so vital to his claim to priority. He spelled out his reason for inquiring about the date of the letter:

> Mr. Wallace who is now exploring New Guinea, has sent me an abstract of the same theory, most curiously coincident even in expressions. And he could never have heard a word of my views. He directed me to forward it to Lyell.– Lyell who is acquainted with my notions consulted with Hooker, (who read a dozen years ago a *long* sketch of mine written in 1844) urged me with much kindness not to let myself to be quite forestalled & to allow them to publish with Wallace's paper an abstract of mine; & as the only very brief thing which I had written out was a copy of my letter to you, I sent it and, I believe, it has just been read, (though never written, & not fit for such purpose) before the Linnean Society.

Despite his victory, Charles Darwin was still preoccupied with Wallace, as he had been since Wallace's first letter arrived at his home on 12 January 1857. He was obsessed with the idea that Wallace, or someone like him, might discover that he had arrived at new thoughts on evolution, because in that crucial letter to Gray the previous summer, he had written:

> You will, perhaps, think it paltry of me, when I ask you not to mention my doctrine; the reason is, if anyone, like the Author of the *Vestiges*, were to hear of them, he might easily work them in, and then I should have to

> quote from a work perhaps despised by naturalists and this would greatly injure any chance of my views being received by those alone whose opinions I value.[2]

Now, only four days after the Linnean meeting, the cause of his concern was no longer vague, but had become clear and specific: it was Alfred Russel Wallace. However it might be explained, nothing about this episode reveals Darwin's character in a good light. However, something then happened that was even more inexplicable, given Lyell and Hooker's evidence to the Linnean Society members.

When Hooker wrote to Wallace shortly after the Linnean meeting to inform him of what had transpired, he included a note from Darwin, who wanted to inform Wallace of the subjects he expected to deal with in the 'Natural Selection' abstract he was then writing. When Wallace received Darwin's letter, he carefully wrote down the chapter headings and their contents in his journal.

Darwin indicated that his projected book would have fourteen chapters, which would deal with the variation of animals in captivity and in their natural habitats, how natural selection worked, the struggle for existence, the laws of variation, the reversion to ancestral type, the gradation of characters, hybridity, instinct, palaeontology and geology, geographical distribution and classification.

There was no reference to Darwin's principle of divergence, or any indication of the importance of divergence in the origination of new species.[3] Yet, only a month before, Darwin had told Hooker that he now considered divergence along with natural selection to be the 'keystone' of his theory.

So why did Darwin not spell out to Wallace that divergence was to be an important part of his 'Natural selection' chapter? Wallace (who had indicated as early as the Sarawak Law that he well understood divergence) would have known that without modification, divergence and extinction, no theory of the origin of species could possibly work. Yet Darwin was seemingly attempting to write his book without referring to this central driving force and regulator.

Darwin's note would have greatly puzzled Wallace, but for Darwin the decision to leave divergence off the list of headings might have seemed expedient. Had Darwin introduced 'Extinction and divergence' as a chapter heading, he might have invited Wallace to write back asking him for his own examples in nature that Darwin could point to which indicated evidence of modification and divergence. That would have been a great embarrassment for Darwin, since he had none. All the work he had carried out (apart from that on barnacles) had been concerned with domestic breeding, in which the millions of years of natural selection, modification, divergence and extinction could never have been replicated.

Of course, when *On the Origin of Species* was published, there, tucked away in the 'Natural selection' chapter, was a whole section under the heading 'Divergence of Character'.

When Wallace eventually returned to England from the Malay Archipelago in 1862, the great controversy Darwin's book had provoked between supporters of evolution and the Church still continued and, through his connection with Darwin, Wallace was to become famous. Although the two men became friendly, Wallace was given no insight into how Darwin's ideas had been influenced by his three fateful letters. Nearly twenty years later, Darwin, who alone knew exactly what had happened, found a way to get the impecunious Wallace a civil pension of £200 a year,[4] but he never again spoke or wrote about the period between 1855 and 1858. After Lyell's letter of May 1856 urging him to publish something to protect his scientific priority, no letters survive in his files from any correspondent who might have sent him information, advice or support in this matter. Until his death, Wallace firmly believed that it was Darwin who had first arrived at the solution to the longstanding species question. He insisted to anyone who would listen that the theory of evolution should be known to everyone as 'Darwinism'.

There were some indications that his work might have received greater consideration, but they did not come from Charles Darwin. Two years after Wallace returned home for good, he received a letter from Sir Charles Lyell.[5] Lyell wrote to compliment him on the clarity and fairness of a recent paper that Wallace had published. The letter concluded with a strange sentiment, given Lyell's behaviour at the Linnean Society meeting six years before: 'The manner in which you have given Darwin the whole credit of the theory of Natural Selection is very handsome, but if anyone else had done it without allusion to your papers it would have been wrong.'

Three years later, Wallace received another letter from Lyell that must have pleased and surprised him in equal measure. Lyell wrote:

> I have been reading over again your paper published in 1855 in the *Annals* on 'The law which has regulated the introduction of new species', passages of which I intend to quote not in reference to your priority of publication but simply because there are some points laid down more clearly than I can find in the work of Darwin itself, in regard to the bearing of the geological and zoological evidence on geographical distribution and the

> Origin of species. I have been looking into Darwin's historical sketch thinking to find some allusion to your essay [...] when he gets to 1855, but I can find no allusion to it. Yet surely I remember somewhere a passage in which Darwin says in print that you had told him in 1855 that you meant by such expressions as 'species being created on the type of pre-existing ones closely allied' and by what you say of modified prototypes, and by the passage in which you ask 'what rudimentary organs mean if each species has been created independently', etc., that new species were created by variation and in the way of ordinary generation.[6]

Such statements by Wallace to Darwin have never come to light, but Lyell's information here seems concrete and specific. Might he have been referring to the contents of Wallace's first letter to Darwin which was not dated 1855 but which arrived in January 1857?

When the third edition of *On the Origin of Species* was published in April 1861, suddenly, and for the first time, Darwin accorded Wallace (in a single sentence) the acknowledgement he deserved as the co-discoverer of the theory of evolution. He probably did not know it was Lyell who indicated to Darwin that this was the right thing to do.

The very first intimation Wallace received that the version of the story he had been given while Darwin was alive might not have been true came with the 1887 publication of a three-volume biography of Darwin by his son Francis. There, Wallace read, for the first time, about the effect his Ternate Law paper had on Darwin. Wallace explained to Francis Darwin:[7]

> I was not aware before that your father had been so distressed – or rather disturbed – by my sending him my essay from Ternate, and I am very glad to feel that his exaggerated sense of honour was quite needless so far as I was concerned, and that the incident did not in any way disturb our friendly relations. I always felt, and still feel, that people generally give me far too much credit for my mere sketch of the theory – so very small an affair as compared with the vast foundation of fact and experiment on which your father worked.

There was something else in the book that might have helped Wallace come to terms with the rejection he had felt when his Sarawak Law received so little attention more than thirty years before. Thomas Huxley, Darwin's most vociferous supporter in the years immediately following the publication of *Origin*, had contributed a personal memoir. Huxley wrote about the influential figures who had helped fend off the anti-transmutationists, and said that no list of names would be complete without mentioning Wallace, whose 'powerful [Sarawak Law] essay ... was published in 1855. On reading it afresh, I have been astonished to recollect how small was the impression it made.'[8]

In 1869 Wallace published his own masterpiece, *The Malay Archipelago*, which cemented his reputation as a naturalist. He dedicated it to Charles Darwin.

Thirty-five years after Darwin's death, Wallace's letters – including those from Lyell and Darwin – were published, and it became known that the letter that had caused Darwin such worry over priority was one of three that he had received from Wallace. Darwin seems to have told none of his contemporaries that he had been well aware of Wallace's revolutionary ideas on species since receiving the first letter, which had been sent from the Malay Archipelago in October 1856.

Charles Darwin was a very secretive man with a driving ambition. He neither praised nor tipped his hat in the direction of Jean-Baptiste Lamarck or of his grandfather Erasmus. He never openly acknowledged his debt to Edward Blyth, nor to Patrick Matthew (who had been one of the first to write about the 'natural means of selection', a phrase that Darwin modified and used without attribution). He never acknowledged his debt to Wallace. By the time Eiseley, Gruber, Beddall, McKinney, Brackman and Brooks began reassembling the long-lost pieces of the jigsaw, the myth-making surrounding Darwin's achievement, which had so worried Darlington in 1959, was complete.

Wallace knew the characteristics of human nature only too well. Interviewed in 1912, a year before his death, he told an American reporter:

> Truth is born into this world only with pangs and tribulations and every fresh truth is received unwillingly. To expect the world to receive a new truth, or even an old truth, without challenging it is to look for one of those miracles which do not occur.[9]

He was speaking, of course, of the immense difficulty they had all experienced in attempting to get the theory of evolution accepted by a hostile world, but he could as well have been talking about how his own insights into the origin of species would still be treated by a complacent scientific establishment a hundred years after his death.

CHAPTER 31

Aftermath

HAD Alfred Russel Wallace sent his letter of March 1858 not to Charles Darwin but to the editor of the *Annals and Magazine of Natural History*, it is likely that we would today talk about Wallaceism rather than Darwinism. It is simply not credible that the Ternate Law would not have been printed by an editor who had already published six articles by Wallace in his magazine over the previous three years (a period during which Darwin had published nothing).

Once published, priority would have been unarguable. The paper, containing just over four thousand words, was so succinct and convincing that there would have been no opportunity for Charles Darwin to argue that he, too, had once read Malthus and written passages in an unpublished work on the subject of the species question. Even his letter to Asa Gray would not, in the light of Gray's apparent reaction to it as 'grievously hypothetical', have carried much weight in such a situation.

Furthermore, had Darwin not received Wallace's letter of October 1856 it is unlikely that he would have understood divergence by the time the Ternate Law arrived, since at the time he was still arguing that organic migration to newly formed environments was a far more convincing explanation of geographical distribution and species change.

Finally, given the well-documented reliability of the mail service, it is hugely unlikely that two out of only three letters from Wallace would have been received by Charles Darwin so much later than their scheduled delivery date.

* * *

The television producer Elin Rhys suggested to me in 1996 that I should examine whether Charles Darwin had acted improperly in order to establish his priority. I felt that she had to be wrong and scoured my mind for what I knew about Darwin. At that time, sad to say, it was not much. I remembered that Darwin had discovered the idea of natural selection after visiting the Galapagos on the *Beagle*. I recalled that he had written out his theory based on the concept of natural selection, and then suppressed it for many years because of his fear that its radical ideas might lead to rioting and social upheaval. I was also aware that Darwin eventually had been forced to publish because he received a letter indicating that Wallace was close to publishing the same idea. I knew that Darwin had written *On the Origin of Species* to ensure that he could claim his place as the man who first formulated the theory of evolution.

Armed with this limited number of 'facts', I defended Darwin's reputation, only to find that Elin was not convinced. What, she wanted to know, did I know of Alfred Russel Wallace? I knew only that he was the man who had written the letter that caused Darwin to act. Did I know that Darwin and two of his friends had cheated Wallace out of his rightful inheritance? Did I know that it was Wallace and not Darwin who had first written out the complete theory?

Stretching my knowledge of Darwin to the absolute limit, I argued that he would not have been capable of such deceit. The BBC *Timewatch* documentary had presented him as a great, kind and gentle man. What made her think he was capable of such duplicity? Two days later, she referred me to Brackman's book *A Delicate Arrangement*, and so began my research into the contentious world of Darwin and Wallace.

Now, I am convinced that Charles Darwin – British national hero, hailed as the greatest naturalist the world has ever known, the originator of one of the greatest ideas of the nineteenth century – lied, cheated and plagiarised in order to be recognised as the man who discovered the theory of evolution.

One question has puzzled me throughout the writing of this book: Why has there been so little attempt by academics to discover when Charles Darwin first heard directly from Alfred Wallace about his ideas on species? Once I began my own research, I realised it was the least studied period of the development of both men's ideas, and perhaps the most productive. So it was relatively easy to concentrate my own inquiries on those few years.

Did Wallace's first letter, written on 10 October 1856, really not arrive at Darwin's home until a few days before the end of April 1857? I needed to know because that period was the most fruitful in terms of the ideas that Darwin was to develop into the crucial chapters of *On the Origin of Species*. Even if it was credible that the letter's delivery had been so delayed, was it conceivable that exactly the same thing should again happen eighteen months later? It seemed to be a tremendously unlikely coincidence that two letters by his only immediate rival in the race to find a solution to the theory of evolution should be delayed in this way.

It was a question others had touched on but not pursued. Brackman had speculated, Beddall had dismissed the first letter without knowing all the circumstances and McKinney got the date of the first letter completely wrong. Few academics found any significance in the serial publication of Wallace's ideas after 1855, and Darwin's unexplained revision of his theory over the same period, so that by late 1856 and early 1857 he was using those same ideas.

Everyone seems to have accepted Darwin's statements around this time as being necessarily honest and truthful. Academics who have studied this period, and read Ospovat, know that the basic tenets of Darwin's longstanding theory, first outlined in 1842, began to change immediately after the publication of Wallace's birds paper in September 1856. Moreover, it is common knowledge among Darwin scholars that Wallace's first letter to Darwin ought to have arrived at Down House four months before Darwin admitted its arrival. Those same academics will probably have read Eiseley's description of how Darwin used Edward Blyth's ideas in his first species notebook, and will also be aware of Gruber's discovery of Darwin's interpolations of his species ideas of 1845 into journal entries he had written aboard the *Beagle* between 1831 and 1836. Darwin, it seems, had a long career of taking credit that was actually due to his contemporaries, or for making dishonest claims about when he had first discovered or understood aspects of the species question. Yet few who have written on this subject seem to have taken a serious forensic interest in when Wallace's letters arrived in London and, therefore, at Darwin's home. How is this to be explained? Why has this crucial period not interested Darwin scholars?

Academics have gone to great lengths to offer explanations of how Darwin developed his new ideas, but none of them involves the original thoughts of Alfred Wallace. Perhaps an examination of Wallace's work in the light of his correspondence with Darwin would force an admission that Darwin had taken these ideas from Wallace. That, as Darlington said in 1959, is not considered desirable. The myth must be respected. Wallace's ideas – which were available to Darwin in magazine articles and personal letters – have been ignored, while simple individual sentences here and there in Darwin's notes have been developed by scholars into a fully-fledged theory of divergence and modification. The general reader is expected to accept this without question. It is not as though these writers even agree about which sentences are significant. It has been suggested that Darwin's understanding of divergence was prompted by his botanical arithmetic, or by a sentence reflecting on his barnacle research, or by new ideas about embryos, or by the practice of the division of labour in his uncle's factory. Yet all the time, Wallace's Sarawak Law and birds article were sitting in the archives.

Scholars and academics have failed to examine this period with anything like sufficient rigour. So, is that the conspiracy at the heart of this book? Well, yes and no.

The sequence of events that drives this story has not been recently uncovered. Most of it has been known to academics and biographers on both sides of the Atlantic for more than forty years. Many of the primary sources (a date-stamped letter, altered manuscript pages on different paper than the original folios, Darwin's first reply to Wallace) have been in the public domain for more than twenty five years. Some of the evidence, distilled from accounts of the lives of both Wallace and Darwin, which appears in this book for the first time, was just waiting to be assessed and critically examined by one of the hundreds of Darwin scholars around the world. Yet in very few of the books about Darwin published during the last forty years will a general reader find a positive reference to any of this evidence. It is a controversial matter, and emotionally fraught for many who consider Darwin to be one of our greatest scientific figures and national heroes.

However, there are two men at the centre of this story, and they have not been treated fairly. Why extol Darwin and diminish the achievement of Wallace? The answer is complicated, but I believe that there is more than a pinch of jealousy involved and a particularly British preoccupation with class.

Although the Linnean Society meeting judged Darwin and Wallace to be equally worthy of recognition as the originators of the theory of evolution, one of them had to be recognised as pre-eminent. The Linnean Society was made up of gentlemen natural philosophers. Wallace, who had written out the complete theory of evolution to which they had listened in silence, was not a gentleman. Charles Darwin, whose unconnected thoughts were contained in two extracts from a 14-year-old essay and a copy of a recent letter, was a gentleman. In the social context of the time, a gentleman always trumped an employee. Thus, the document merging the two presentations referred to the 'Darwin–Wallace' theory of evolution. In the lifetimes of both men (Darwin died in 1881, Wallace in 1913) it was usual for the theory to be referred to by this title, but after Wallace's death it became 'Darwin's theory of evolution'. For almost a century, Wallace's scientific achievement has been effectively buried under Darwin's reputation.

Wallace, who until late in his life had no inkling of the scandalous background to events at the Linnean Society, had not endeared himself to the establishment by arguing for unpopular causes and for some outlandish beliefs (such as mesmerism and spiritualism). He argued vehemently against the 'barbarism' shown towards the Australian Aborigines by English settlers and questioned the 'civilising influence' of Christian missionaries in the northern Celebes. He also supported land reform in Britain, arguing that without a more equitable distribution of land, it would not be possible to have a just society. A mixture of snobbery and ridicule diminished his reputation. Darwin's reputation, on the other hand, has increased dramatically as the power of organised religion has declined.

In the course of researching this book, I have been told that academics today are not interested in what happened one hundred and fifty years ago; that what happened at the Linnean Society meeting was 'naughty', as though we were discussing the transgressions of a child; and that if I persisted, I would need to get a steel helmet and dig a deep hole. These comments were made by academics involved in Darwin studies, the history of science and evolutionary biology. One academic, on hearing the gist of this story but not its title, said that he would strongly resist a suggestion that there was any conspiracy among his colleagues to favour Darwin above Wallace.

Whether the title of this book is apposite or merely contentious is something others will have to decide. However, the fact that a detailed narrative of these events has not been made accessible to the general reader long before now says a lot about our need to maintain convenient myths and beliefs that were established in the distant past. This need has been commented on by some academics, as the quotations at the beginning of this book attest.

The fact that every research project that has attempted to uncover the truth of this story was an American initiative, rather than British, also says a great deal about the difference in openness between the two countries. If it had been down to academics from the United Kingdom, I suspect we would know very little about the thwarted genius of Alfred Russel Wallace, and probably absolutely nothing about the dark and desperate side of Charles Darwin.

The modern researchers who looked seriously at this subject are all now dead. Their work, and that of Wallace, deserves a better fate than to be continually bypassed in the unseemly rush to burnish Darwin's image. Recently, more books about Wallace have been published. Almost all include a version of his famous letter to Darwin, and most repeat dutifully that it was received on 18 June 1858. Few hail the paper enclosed with that letter as a masterpiece in its own right, worthy of securing absolute priority for its author.

Still, Wallace has undisputed claim to being the man who first gave the world the answer to the problem of the origin of species. The publication of his Ternate Law paper in August 1858 in the journal of the Linnean Society pre-dated the publication of Darwin's *Origin* by fifteen months. Yet it is still scant reward for a brilliant yet unassuming naturalist who was never to comprehend the full extent of the conspiracy enacted against him.

Cast of Characters

Darwin's Contemporaries

Henry Walter Bates (1825–1892) Naturalist and entomologist. Wallace's companion and friend. Travelled with Wallace to the Amazon basin in 1848 with the shared objective of finding a solution to what was then known as the 'species question'.

Edward Blyth (1810–1873) Blyth's articles on natural selection were published just before Darwin opened his first scientific notebook. Thought by Loren Eiseley to have directly influenced Darwin's early ideas on transmutation and natural selection. Mentioned by Darwin in *On the Origin of Species* for his classification skills, but not as a theorist.

Edward Forbes (1815–1854) Geologist and friend of Darwin and Lyell who suggested that oceanic islands were formerly areas of nearby continents split off by immense geological forces over unknowable periods of time. His insistence that such islands retained species identical to those found on the mainland caused Darwin great consternation.

Asa Gray (1810–1888) Harvard botanist to whom Darwin wrote in confidence that he had discovered a principle of divergence. The letter became a crucial piece of evidence at the Linnean meeting in July 1858, as it was claimed that it indicated that Darwin had an understanding of divergence before receiving Wallace's letter in June 1858.

Joseph Hooker (1817–1911) Botanist. Lifelong friend and confidant whose knowledge and expertise guided Darwin as he developed his ideas. Hooker, without question or hesitation and in close collaboration with Lyell, helped Darwin achieve scientific priority on the species question at the Linnean meeting.

James Hutton (1726–1797) Considered by many to be the father of modern geology, Hutton made several important discoveries about the relative ages of the rocks he studied in his native Scotland. Hutton's ideas were not widely understood until Lyell's restatement of them in the 1830s.

Jean-Baptiste Lamarck (1744–1829) Entomologist and zoologist. Proposed a theory of evolution based on the idea that during its lifetime, an animal could acquire characteristics that gave it an advantage in the struggle for survival, and which could be passed on to the next and succeeding generations.

Sir Charles Lyell (1797–1875) Geologist, and Darwin's friend and mentor. His *Principles of Geology*, which proposed that geological change is the result of a steady accumulation of minute changes over vast spans of time, also contained the idea that new species are created by God as and when they are required to maintain the harmony of a world existing in a state of perfect adaptation.

Patrick Matthew (1790–1874) Fruit-grower. Wrote a book on naval timber and arboriculture in 1831 (the year Darwin joined the *Beagle*), in which he discussed the idea of 'natural means of selection' while referring to species change. His claim that Darwin took and modified his phrase to 'natural selection' while using his ideas without attribution was rebuffed by Darwin.

Thomas Wollaston (1822–1878) Entomologist. Correspondent who first offered Darwin evidence that similar species of land molluscs and wingless insects found on both Madeira and the coast of Africa gave proof that Edward Forbes's land bridges idea had merit.

The Modern Researchers

Barbara G. Beddall. Zoologist and early champion of the theory of continental drift, with its implications for the geographical distribution of species. Frustrated by failing to find crucial letters from Wallace and others among the archives at Cambridge, she concluded that they had been intentionally removed to deliberately obscure the story of how Darwin arrived at his theory.

Arnold C. Brackman. Writer and journalist who first published an account of some of this story of plagiarism in his book *A Delicate Arrangement* (1980).

John Langdon Brooks. Environmental biologist who could not believe that the similarities between the accounts of divergence in the papers of Darwin and Wallace were coincidences. Proved to himself that Wallace's March 1858 letter from Ternate had been received by Darwin earlier than Darwin had claimed.

Cyril Dean Darlington. Sherardian Professor of Botany at Oxford University. Asked how Darwin had managed to arrive at the theory of evolution without leaving any clues about his thought processes. Suggested that Darwin had simply edited together other people's ideas, principally those of Lyell and Hooker.

Loren Eiseley. Anthropologist and writer. Argued that Darwin first conceived of his species ideas after reading articles written by Edward Blyth a whole year before he came across the ideas of Thomas Malthus. Decided the direct influence was Blyth and not Lyell on the basis of the word '*inosculate*', used by Blyth in his articles but never previously used by Darwin.

Howard E. Gruber. Cognitive psychologist. His comparative analysis of both editions of Darwin's *Beagle* journals revealed that in 1845 Darwin had made significant additions to his original account without explanation. These changes gave the impression that while still on the voyage, he had ideas about the species question which he could only have gleaned after years of expert consultation and theoretical insight.

H. Lewis McKinney. Historian of science. Studied Wallace's notebooks at the Linnean Library. Discovered Darwin's original copy of Wallace's Sarawak Law with crucial annotations, and helped to uncover Darwin's deceit concerning the date on which Wallace's final letter arrived.

Dov Ospovat. Historian of science. After minute examination of Darwin's papers at Cambridge, Ospovat concluded that Darwin's conception of natural selection in his essay of 1844 was entirely different from that outlined in *On the Origin of Species* fifteen years later. Moreover, he discovered that Darwin had not understood the idea of divergence until sometime in the last six months of 1856.

Leonard G. Wilson. Historian. Discovered notebooks of Lyell's that others had missed. The notebooks indicated that in November 1855, Lyell had read Wallace's Sarawak Law, had been hugely impressed and soon afterwards advised Darwin to publish his ideas or risk losing scientific priority.

Glossary

Atlantis theory (also 'land bridges' or 'continental extensions') The idea that oceanic islands are areas of land that were once joined to the closest continental land mass before being separated by geological forces.

Coleoptera The largest order of insects containing beetles and weevils.

Common descent The idea that all organisms are ultimately descended from a common ancestor.

Divergence Modification and descent over time of one species into closely allied species.

Entomology The study of insects.

Evolution (descent with modification) The gradual accumulation of small changes in organisms over many generations leading to the formation of new species.

Family A group of animals or plants of similar genera.

Genus (genera) A group of animals or plants of similar species.

Geographical distribution The ranges of particular plants and animals; the study of where they are and why.

Lamarckism The theory of evolution proposed by Lamarck in which characteristics acquired by habit or will in one generation of a species can be inherited by the next, leading to the gradual formation of new species.

Lepidoptera Insect order containing butterflies and moths.

Malthus's 'principle of population' The power of population to increase is infinitely greater than the power of the earth to produce subsistence for man. Population when unchecked increases in a geometrical ratio: subsistence increases only in an arithmetic ratio.

Modification The process of gradual change from one species into another.

Natural selection Selection by the environment. Those individuals better adapted to an environment are more likely to survive and reproduce their kind. Continued generation after generation, the inheritance of small differences or changes leads to the formation of new species.

Ornithology The study of birds.

Sarawak Law 'Every species has come into existence coincident both in space and time with a pre-existing closely allied species.' (Wallace, 1855.)

Species An interbreeding group of organisms. At any one time, species are separated from one another by various barriers to interbreeding. Considered over time, however, each one is related to its immediate predecessor by descent.

Species question The conundrum of how new species arise. A problem that exercised many of the best scientific minds during the period covered by this book.

Survival of the fittest Success in the struggle for continuing existence during times of environmental stress. A phrase coined by Herbert Spencer in 1852.

Ternate Law The tendency in nature for varieties to progress indefinitely away from the original type. (Wallace, 1858.)

Transmutation of species The gradual change of one species into another.

Tropical rainforest Dense, luxuriant forest in regions where rainfall averages from under two hundred and fifty centimetres to more than two thousand centimetres per year, and where temperature averages about 27°C.

Variation Inherited differences between individuals of a species.

Variety (subspecies) Individuals within a species that can cross-breed freely with other varieties if brought into contact. Varieties will usually be geographically separated from each other.

Wallace Line An imaginary line drawn between the islands of Bali and Lombok, between Borneo and Celebes and south of the Philippines, which marks the division of the Malay Archipelago into two parts. To the east of the line the animal population represents the Australian biota, and to the west, the Asian biota.

References

Pearsall, Judy, 1999, *Concise Oxford Dictionary* (tenth edition), Oxford, Oxford University Press.

Beddall, Barbara G., 1969, *Wallace and Bates in the Tropics*, London, Macmillan.

APPENDIX 1

Darwin and Wallace: timeline of ideas, 1831–1862

Date	*Darwin*	*Wallace*
1831	Begins *Beagle* voyage.	At school in Hertfordshire.
1836	*Beagle* voyage ends in October. Charles Lyell becomes a great friend.	
1837	Opens transmutation notebooks. Discovers idea of natural selection.	Leaves school at age 14. Assistant surveyor in Wales. Interest in natural history develops.
1839	Publishes first account of *Beagle* voyage.	Meets H. W. Bates in Leicester. Learns about beetles. Reads accounts of explorers and travellers.
	Ridicules *Vestiges* and dismisses its ideas of species change.	Enthuses about *Vestiges* to Bates.
1842	Writes 30-page sketch of idea that. migration to oceanic islands accounts for new species.	Suggests to Bates they become professional collectors with view to solving the origin of species.
1844	Expands sketch to 230-page essay.	
1845	Second edition of *Beagle* voyage appears. Galapagos ideas appear for first time.	
1846	Edward Forbes argues that oceanic islands formerly part of continent. Hooker unimpressed by 1844 essay. Darwin begins scientific study of barnacles.	
1847	Hooker criticises Darwin's species migration idea.	
1848		Wallace and Bates leave for Brazil.
1852		Wallace loses collections in ship fire.
1853		Wallace tells Society of Entomologists that species mutate. Believes that there is some principle in nature that regulates species.

1854	Darwin finishes barnacles study and begins again on species. Still obsessed with Forbes's Atlantis theory. Begins salt-water experiments to show migration of plants possible.	Wallace leaves for the Malay Archipelago.
1855	Still believes new species are formed only on oceanic islands. No variation in environments unaffected by geological change.	Wallace publishes his Sarawak Law: all species come into existence closely related, in both time and space, to pre-existing species. Species have diverged from a common ancestor.
1856	In April, Charles Lyell sees Wallace as a threat. Darwin begins his 'big species book' – with the working title 'Natural Selection' – to forestall Wallace. July: Friends accept Forbes's ideas, but Darwin still continues with seed experiments. September: following publication of Wallace's birds article, Darwin becomes very interested in geographical distribution of species.	May: Discovers line separating Australian fauna from Asian. Realises that migration theory cannot account for huge differences in animal species on islands so close together. Wallace publishes two more articles in July and September. He states that humans and orang-utans probably evolved from the same ancestor, and shows how species originate from common ancestral species. October: writes to Darwin about his discovery. Outlines ideas on species.
1857	January to March: Wallace's letter arrives in January 1857. Experiments with seeds and pigeons no longer so important. Now admits Forbes's Atlantis theory might work. The last day of March: adds a 'principle of divergence' to 'Natural Selection' manuscript. April 12: tells Hooker of sudden insight that varieties and new species are the same thing. May: writes to Wallace saying letter reached him 'a few days ago', and that they obviously thought much alike. 5 September: outlines new divergence idea to Asa Gray. Keeps a copy of the outline. Tacks 'Principle of divergence' onto old ideas. No explanation of how it works, but considers it very important for explaining origin of varieties and species. December: receives Wallace's second letter. Replies two weeks later. Rejects Wallace's evidence, which challenges his own migration theory.	January to March: discovers variant butterfly. Claims that Lyell was wrong to argue that species were specially created and perfectly adapted for their environment. April: wonders whether man is a species or a collection of different varieties. Notices how tribes that organise food and water are more likely to survive than those that don't. Summer: sends three accounts of new ideas to London: why Lyell was wrong about species, discovery of a new variety of butterfly and ideas about species and varieties. 27 September: replies to Darwin. Includes new ideas about the geographical distribution of species. Supports Forbes and the oceanic islands theory.

1858		February: while on Gilolo, Wallace realises Malthus's theory is the key to origin of species. Writes his theory between bouts of sickness.
	March: reply to Wallace arrives at Ternate on 9 March. In it, he indicates that both Lyell and Edward Blyth commended to him Wallace's Sarawak Law.	9 March: Sends letter to Darwin.
	June: Wallace letter arrives in London on 3 June. Darwin claims it arrived on 18 June, just after he has rewritten crucial parts of his manuscript. Sends Wallace's letter to Lyell.	
	1 July: Linnean Society meeting. Charles Darwin is granted priority. Wallace named co-discoverer.	
		August: his theory of evolution published more than a year before *Origin.*
		September: discovers from Hooker what had happened on 1 July.
1859	November: Publication of *On the Origin of Species.*	
1862		Wallace returns to England

APPENDIX 2

Southampton–Malay Archipelago Shipping Timetable 1857–58

	Movement of mails	*Dates of movement*				*Sources*
1	Mails depart Southampton	Sept. 20 1857 *Ripon*	Oct. 20 1857 *Indus*	Nov. 20 1857 *Nemesis*	Dec. 20 1857 *Pera*	*Lloyd's List, The Times*
2	Arrive at Singapore	Nov. 1 1857 *Singapore*	Dec. 4 1857 *Cadiz*	Jan. 2 1858 *Ottawa*	Feb. 1 1858 *Singapore*	*Singapore Free Press*
3	Depart Singapore for Batavia	Nov. 2 1857 *Banda*	Dec. 5 1857 *Banda*	Jan. 3 1858 *Banda*	Feb. 2 1858 *Banda*	*Singapore Free Press*
4	Arrive in Batavia	Nov. 6 1857 *Banda*	Dec. 9 1857 *Banda*	Jan. 8 1858 *Banda*	Feb. 5 1858 *Banda*	*Javasche Courant*
5	Depart Batavia for Surabaya	Nov. 7 1857 *Padang*	Dec. 10 1857 *Soerabaja*	Jan. 9 1858 *Makassar*	Feb. 7 1858 *Banda*	*Javasche Courant*
6	Arrive at Surabaya	Nov. 10 1857 *Padang*	Dec. 13 1857 *Soerabaja*	Jan. 12 1858 *Makassar*	Feb. 10 1858 *Banda*	*Javasche Courant*
7	Depart Surabaya for the Moluccas via Macassar and Timor	Nov. 14 1857 *Padang*	Dec. 16 1857 *Ambon*	Jan. 17 1858 *Makassar*	Feb. 14 1858 *Ambon*	*Javasche Courant*
8	Depart Macassar	Nov. 19 1857 *Padang*	Dec. 20 1857* *Ambon*	Jan. 21 1858* *Makassar*	Feb. 19 1858* *Ambon*	Wallace's journal
9	Arrive at Amboina	Nov. 30 1857 *Padang*	Jan. 4 1858 *Ambon*	– *Makassar*	– *Ambon*	Based on Wallace's journal
10	Arrive at Ternate	– *Padang*	Jan. 8 1858 *Ambon*	– *Makassar*	Mar. 9 1858 *Ambon*	Based on Wallace's journal
11	Depart Macassar for Surabaya	Dec. 18 1857 *Padang*	Jan. 22 1858 *Ambon*	Feb. 18 1858 *Makassar*	Mar. 19 1858 *Ambon*	*Javasche Courant*
12	Arrive at Surabaya	Dec. 25 1857 *Padang*	Jan. 26 1858 *Ambon*	Feb. 22 1858 *Makassar*	Mar. 24 1858 *Ambon*	*Javasche Courant*
13	Depart Surabaya for Batavia	Jan. 3 1858 *Koningin der Nederlanden*	Feb. 3 1858 *Palembang*	Feb 25 1858 *Makassar*	Mar. 26 1858 *Ambon*	*Javasche Courant*
14	Arrive at Batavia	Jan. 7 1858 *Koningin der Nederlanden*	Feb. 8 1858 *Palembang*	Mar. 2 1858 *Makassar*	Mar. 31 1858 *Ambon*	*Javasche Courant*
15	Depart Batavia for Singapore	Jan. 12 1858 *Koningin der Nederlanden*	Feb. 12 1858 *Koningin der Nederlanden*	Mar.12 1858 *Koningin der Nederlanden*	April 12 1858 *Koningin der Nederlanden*	*Singapore Free Press*
16	Depart Singapore for Suez	Jan. 23 1858 *Aden*	Feb. 23 1858 *Ganges*	Mar. 23 1858 *Norna*	April 21 1858 *Bombay*	*Singapore Free Press*
17	Second Class Mails arrive in Southampton & transferred to London	Mar. 5 1858 *Pera*	April 3 1858 *Indus*	May 10 1858 *Colombo*	June 2 1858 *Pera*	*The Times*
18	Letters delivered in London	Mar. 6 1858	April 4 1858	May 11 1858	June 3 1858	Post Office records

* estimated departure times based on Wallace's entries

Sources: Numbers refer to row in the table

1) Mails depart Southampton
Lloyd's List
No 13,564, 21 September 1857, pg 2.
The Times
No 22,819, 23 October 1857, pg 5;
No 22,844, 21 November 1857, pg 7;
No 22,869, 21December 1857, pg 4.

2) Arrive at Singapore
Singapore Free Press
Vol 24, No 48, 19 November 1857, pg 3;
Vol 24, No 51, 10 December 1857, pg 3;
Vol 25, No 1, 7 January 1858, pg 3;
Vol 25, No 5, 4 February 1858, pg 3.

3) Depart Singapore for Batavia
Singapore Free Press
Vol 24, No 46, 5 November 1857, pg 3;
Vol 24, No 51, 10 December 1857, pg 3;
Vol 25, No 1, 7 January 1858, pg 3;
Vol 25, No 5, 4 February 1858, pg 3.

4) Arrive in Batavia
Javasche Courant
No 91, 11 November 1857;
No 99, 12 December 1857;
No 3, 9 January 1858;
No 12, 10 February 1858.

5) Depart Batavia for Surabaya
Javasche Courant
No 90, 11 November 1857;
No 99, 12 December 1857;
No 4, 13 January 1858;
No 12, 10 February 1858.

6) Arrive at Surabaya
Javasche Courant
No 93, 21 November 1857;
No 101, 19 December 1857;
No 7, 23 January 1858;
No 15, 20 February 1858.

7) Depart Surabaya for the Moluccas via Macassar and Timor
Javasche Courant
No 94, 25 November 1857;
No 103, 26 December 1858;
No 8, 27 January 1858;
No 16, 24 February 1858.

11) Depart Macassar for Surabaya
Javasche Courant
No 2, 6 January 1858;
No 11, 6 February 1858;
No 18, 3 March 1858;
No 27, 3 April 1858.

12) Arrive at Surabaya
Javasche Courant
No 2, 6 January 1858;
No 11, 6 February 1858;
No 18, 3 March 1858;
No 27, 3 April 1858.

13) Depart Surabaya for Batavia
Javasche Courant
No 3, 9 January 1858;
No 12, 10 February 1858;
No 19, 6 March 1858;
No 27, 3 April 1858.

14) Arrive at Batavia
Javasche Courant
No 3, 9 January 1858;
No 12, 10 February 1858;
No 19, 6 March 1858;
No 27, 3 April 1858.

15) Depart Batavia for Singapore
Singapore Free Press
Vol 25, No 3, 21 January 1858, pg 3;
Vol 25, No 7, 18 February 1858, pg 3;
Vol 25, No 11 18 March 1858, pg 3;
Vol 25, No 16 22 April 1858, pg 3.

16) Depart Singapore for London
Singapore Free Press
Vol 25, No 4, 28 January 1858, pg 3;
Vol 25, No 8, 25 February 1858, pg 3;
Vol 25, No 12, 25 March 1858, pg 3;
Vol 25, No 16, 22 April 1858, pg 3.

17) Second Class Mails arrive in Southampton & transferred to London
The Times
No 22,934, 6 March 1858, pg 10;
No 22,959, 5 April 1858, pg 10;
No 22,990, 11 May 1858, pg 11;
No 23,010, 3 June 1858, pg 9.

18) Letters delivered in London
Indian & Australian Mail Homeward, 1857-61(Post 43/156), Post Office Museum, Mount Pleasant, London.

APPENDIX 3

Route map of mail steamers in the Dutch East Indies in the late 1850s

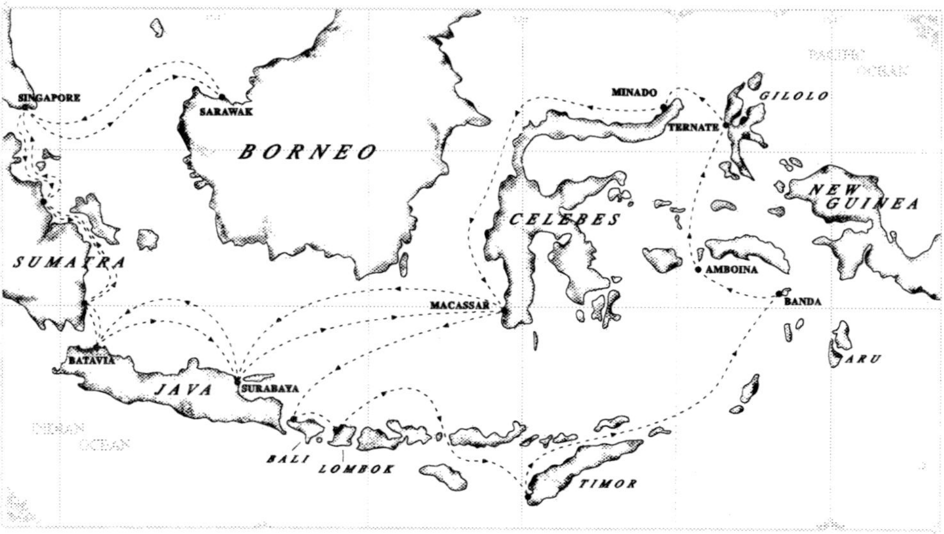

APPENDIX 4

On the Tendency of Varieties to depart indefinitely from the Original Type
By Alfred Russel Wallace

One of the strongest arguments which have been adduced to prove the original and permanent distinctness of species is, that varieties produced in a state of domesticity are more or less unstable, and often have a tendency, if left to themselves, to return to the normal form of the parent species; and this instability is considered to be a distinctive peculiarity of all varieties, even of those occurring among wild animals in a state of nature, and to constitute a provision for preserving unchanged the originally created distinct species.

In the absence or scarcity of facts and observations as to varieties occurring among wild animals, this argument has had great weight with naturalists, and has led to a very general and somewhat prejudiced belief in the stability of species. Equally general, however, is the belief in what are called 'permanent or true varieties,' – races of animals which continually propagate their like, but which differ so slightly (although constantly) from some other race, that the one is considered to be a variety of the other. Which is the variety and which the original species, there is generally no means of determining, except in those rare cases in which one race has been known to produce an offspring unlike itself and resembling the other. This, however, would seem quite incompatible with the 'permanent invariability of species,' but the difficulty is overcome by assuming that such varieties have strict limits, and can never again vary further from the original type, although they may return to it, which, from the analogy of the domesticated animals, is considered to be highly probable, if not certainly proved.

It will be observed that this argument rests entirely on the assumption, that varieties occurring in a state of nature are in all respects analogous to or even identical with those of domestic animals, and are governed by the same laws as regards their permanence of further variation. But this is the object of the present paper to show that this assumption is altogether false, that there is a general principle in nature which will cause many varieties to survive the parent species, and to give rise to successive variations departing further and further from the original type, and which also produces, in domesticated animals, the tendency of varieties to return to the parent form.

The life of wild animals is a struggle for existence. The full exertion of all their faculties and all their energies is required to preserve their own existence and provide for that of their infant offspring. The possibility of procuring food during the least favourable seasons, and of escaping the attacks of their most dangerous enemies, are the primary conditions which determine the existence both of individuals and of entire species. These conditions will also determine the population of a species; and by a careful consideration of all the circumstances we may be enabled to comprehend, and in some degree to explain, what at first sight appears so inexplicable – the excessive abundance of some species, while others closely allied to them are very rare.

The general proportion that must obtain between certain groups of animals is readily seen. Large animals cannot be so abundant as small ones; the carnivora must be less numerous than the herbivora; eagles and lions can never be so plentiful as pigeons and antelopes; the wild asses of the Tartarian deserts cannot equal in numbers the horses of the more luxuriant prairies and pampas of America. The greater or less fecundity of an animal is often considered to be one of the chief causes of its abundance or scarcity; but a consideration of these facts will show us that it really has little or nothing to do with the matter. Even the least prolific of animals would increase rapidly if unchecked, whereas it is evident that the animal population of the globe must be stationary, or perhaps, through the influence of man, decreasing. Fluctuations there may be; but permanent increase, except in restricted localities, is almost impossible. For example, our own observation must convince us that birds do not go on increasing every year in a geometrical ratio, as they would do, were there not some powerful check to their natural increase. Very few birds produce less than two young ones each year, while many have six, eight, or ten; four will certainly be below the average; and if we suppose that each pair produce young only four times in their life, that will also be below the average, supposing them not to die either by violence or want of food. Yet at this rate how tremendous would be the increase in a few years from a single pair! A simple calculation will show that in fifteen years each pair of birds would have increased to nearly ten millions! whereas we have no reason to believe that the number of birds of any country increases at all in fifteen or in one hundred and fifty years. With such powers of increase the population must have reached its limits, and have become stationary, in a very few years after the origin of each species. It is evident, therefore, that each year an immense number of birds must perish – as many in fact as are born; and as in the lowest calculation the progeny are each year twice as numerous as their parents, it follows that, whatever be the average number of individuals existing in any given country, twice that number must perish annually, – a striking result, but one which seems at least highly probable, and is perhaps under rather than over the truth. It would therefore appear that, as far as the continuance of the species and the keeping up the average number of individuals are concerned, large broods are superfluous. On the average all above one become food for hawks and kites, wild cats and weasels, or perish of cold and hunger as winter comes on. This is strikingly proved by the case of particular species; for we find that their abundance in individuals bears no relation whatever to their fertility in producing offspring. Perhaps the most remarkable instance of an immense bird population is that of the passenger pigeon of the United States, which lays only one, or at most two eggs, and is said to rear generally but one young one. Why is this bird so extraordinarily abundant, while others producing two or three times as many young are much less plentiful? The explanation is not difficult. The food most congenial to this species, and on which it thrives best, is abundantly distributed over a very extensive region, offering such differences of soil and climate, that in one part or another of the area the supply never fails. The bird is capable of a very rapid and long-continued flight, so that it can pass without fatigue over the whole of the district it inhabits, and as soon as the supply of food begins to fail in one place is able to discover a fresh feeding-ground. This example strikingly shows us that the procuring of a constant supply of wholesome food is almost the sole condition requisite for ensuring the rapid increase of a given species, since neither the limited fecundity, nor the unrestricted attacks of birds of prey and of man are here sufficient to check it. In no other birds are these peculiar circumstances so strikingly combined. Either their food is more liable to failure, or they have not sufficient power of wing to search for it over an extensive area, or during some season of the year it becomes very scarce, and less wholesome substitutes have to be found; and thus, though more fertile in offspring, they can never increase beyond the supply of food in the least favourable seasons. Many birds can only exist by migrating, when their food becomes scarce, to regions possessing a milder, or at least a different climate, though, as these migrating birds are seldom excessively abundant, it is evident that the countries they visit are still deficient in a constant and abundant supply of wholesome food. Those whose organization does not permit them to migrate when their food becomes periodically scarce, can never attain a large population. This is probably the reason why woodpeckers are scarce with us, while in the tropics they are among the most abundant of solitary birds. Thus the house sparrow is more abundant than the redbreast, because its food is more constant and plentiful, – seeds of grasses being preserved during the winter, and our farm-yards and stubble-fields furnishing an almost inexhaustible supply. Why, as a general rule, are aquatic, and especially sea birds, very numerous in individuals? Not because they are more prolific than the others, generally the contrary; but because their food never fails, the sea-shores and river-banks daily swarming with a fresh supply of small mollusca and crustacea. Exactly the same law applies to mammals. Wild cats are prolific and have few enemies; why then are they never as abundant as rabbits? The only intelligible answer is, that their supply of food is more precarious. It appears evident, therefore, that so long as a country remains physically unchanged, the numbers of its animal population cannot materially increase. If one species does so, some others requiring the same kind of food must diminish in proportion. The numbers that die annually must be immense; and as the individual existence of each animal depends upon itself, those that must die must be the weakest – the very young, the aged, and the diseased, – while those that prolong their existence can only be the most perfect in health and vigour – those who are best able to obtain food regularly, and avoid their numerous enemies. It is, as we commenced by remarking, 'a struggle for existence,' in which the weakest and least perfectly organized must always succumb.

Now it is clear that what takes place among the individuals of a species must also occur among the several allied species of a group, – *viz.* that those which are best adapted to obtain a regular supply of food, and to defend themselves against the attacks of their enemies and the vicissitudes of the seasons must necessarily obtain and preserve a superiority in population; while those species which from some defect of power or organization are the least capable of counteracting the vicissitudes of food supply, &c., must diminish in numbers, and, in extreme cases, become altogether extinct. Between these extremes the species will present various degrees of capacity for ensuring the means of preserving life; and it is thus we account for the abundance or rarity of species. Our ignorance will generally prevent us from accurately tracing the effects to their causes; but could we become perfectly acquainted with the organization and habits of the various species of animals, and could we measure the capacity of each for performing the different acts necessary to its safety and existence under all the varying circumstances by which it is surrounded, we might be able even to calculate the proportionate abundance of individuals which is the necessary result.

If now we have succeeded in establishing these two points – 1st , that the animal population of a country is generally stationary, being kept down by a periodical deficiency of food, and other checks; and, 2nd , that the comparative abundance or scarcity of the individuals of the several species is entirely due to their organization and resulting habits, which, rendering it more difficult to procure a regular supply of food and to provide for their personal safety in some cases than in others, can only be balanced by a difference in the population which have to exist in a given area – we shall be in a condition to proceed to the consideration of varieties, to which the preceding remarks have a direct and very important application.

Most or perhaps all the variations from the typical form of a species must have some definable effect, however slight, on the habits or capacities of the individuals. Even a change of colour might, by rendering them more or less distinguishable, affect their safety; a greater or less development of hair might modify their habits. More important changes, such as an increase in the power or dimensions of the limbs or any of the external organs, would more or less affect their mode of procuring food or the range of country which they inhabit. It is also evident that most changes would affect, either favourably or adversely, the powers of prolonging existence. An antelope with shorter or weaker legs must necessarily suffer more from the attacks of the feline carnivora; the passenger pigeon with less powerful wings would sooner or later be affected in its powers of procuring a regular supply of food; and in both cases the result must necessarily be a diminution of the population of the modified species. If, on the other hand, any species should produce a variety having slightly increased powers of preserving existence, that variety must inevitably in time acquire a superiority in numbers. These results must follow as surely as old age, intemperance, or scarcity of food produces an increased mortality. In both cases there may be many individual exceptions; but on the average the rule will invariably be found to hold good. All varieties will therefore fall into two classes – those which under the same conditions would never reach the population of the parent species, and those which would in time obtain and keep a numerical superiority. Now let some alteration of physical conditions occur in the district – a long period of drought, a destruction of vegetation by locusts, the irruption of some new carnivorous animal seeking 'pastures new' – any change in fact tending to render existence more difficult to the species in question, and taking its utmost powers to avoid complete extermination; it is evident that, of all the individuals composing the species, those forming the least numerous and most feebly organized variety would suffer first, and, were the pressure severe, must soon become extinct. The same causes continuing in action, the parent species would next suffer, would gradually diminish in numbers, and with a recurrence of similar unfavourable conditions might also become extinct. The superior variety would then alone remain, and on a return to favourable circumstances would rapidly increase in numbers and occupy the place of the extinct species and variety.

The variety would now have replaced the species, of which it would be a more perfectly developed and more highly organized form. It would be in all respects better adapted to secure its safety, ad to prolong its individual existence and that of the race. Such a variety could not return to the original form; for that form is an inferior one, and could never compete with it for existence. Granted, therefore, a 'tendency' to reproduce the original type of species, still the variety must ever remain preponderant in numbers, and under adverse physical conditions again alone survive. But this new, improved, and populous race might itself, in course of time, give rise to new varieties, exhibiting several diverging modifications of form, any of which, tending to increase the facilities for preserving existence, must, by the same general law, in their turn become predominant. Here, then, we have progression and continued divergence deduced from the general laws which regulate the existence of animals in a state of nature, and from the undisputed fact that varieties do frequently occur. It is not, however, contended that this result would be invariable; a change of physical conditions in the district might at times materially modify it, rendering the race which had been the most capable of supporting existence under the former conditions now the least so, and even causing the extinction of the newer and, for a time, superior race, while the old or parent species and its first inferior varieties continued to flourish. Variations in unimportant parts might also occur, having no perceptible effect of the life-preserving powers; and the varieties so furnished might run a course parallel with the parent species, either giving rise to further variations or returning to the former type. All we argue for is, that certain varieties have a tendency to maintain their existence longer than the original species, and this tendency must make itself felt; for though the doctrine of chances or averages can never be trusted to on a limited scale, yet, if applied to high numbers, the results come nearer to what theory demands, and, as we approach to an infinity of examples, becomes strictly accurate. Now the scale on which nature works is so vast – the numbers of individuals and periods of time with which she deals approach so near to infinity, that any cause, however slight, and however liable to be veiled and counteracted by accidental circumstances, must in the end produce its full legitimate results.

Let us now turn to domesticated animals, and inquire how varieties produced among them are affected by the principles here enunciated. The essential difference in the condition of wild and domestic animals is this – that among the former, their well-being and very existence depend upon the full exercise and healthy condition of all their senses and physical powers, whereas, among the latter, these are only partially exercised, and in some cases are absolutely unused. A wild animal has to search, and often to labour, for every mouthful of food – to exercise sight, hearing, and smell in seeking it, and in avoiding dangers, in procuring shelter from the inclemency of the seasons, and in providing for the subsistence and safety of its offspring. There is no muscle of its body that is not called into daily and hourly activity; there is no sense or faculty that is not strengthened by continual exercise. The domestic animal, on the other hand, has food provided for it, is sheltered, and often confined, to guard against the vicissitudes of the seasons, is carefully secured from the attacks of its natural enemies, and seldom even rears its young without human assistance. Half of its senses and faculties are quite useless; and the other half are but occasionally called into feeble exercise, while even its muscular system is only irregularly called into action.

Now when a variety of such an animal occurs, having increased power or capacity in any organ or sense, such increase is totally useless, is never called into action, and may even exist without the animal ever becoming aware of it. In the wild animal, on the contrary, all its faculties and powers being brought into full action for the necessities of existence, any increase becomes immediately available, is strengthened by exercise, and must even slightly modify the food, the habits, and the whole economy of the race. It creates as it were a new animal, one of superior powers, and which will necessarily increase in numbers and outlive those inferior to it.

Again, in the domesticated animal all variations have an equal chance of continuance; and those which would decidedly render a wild animal unable to compete with its fellows and continue its existence are no disadvantaged whatever in a state of domesticity. Our quickly fattening pigs, short-legged sheep, pouter pigeons, and poodle dogs could never have come into existence in a state of nature, because the very first step towards such inferior forms would have led to rapid extinction of the race; still less could they now exist in competition with their wild allies. The great speed but slight endurance of the race horse, the unwieldy strength of the ploughman's team, would both be useless in a state of nature. If turned wild on the pampas, such animals would probably soon become extinct, or under favourable circumstances might each lose those extreme qualities which would never be called into action, and in a few generations would revert to a common type, which must be that in which the various powers and faculties are so proportioned to each other as to be best adapted to procure food and secure safety, – that in which by the full exercise of every part of his organization the animal can alone continue to live. Domestic varieties, when turned wild, must return to something near the type of the original wild stock, or become altogether extinct.

We see, then, that no inferences as to varieties in a state of nature can be deduced from the observation of those occurring among domestic animals. The two are so much opposed to each other in every circumstance of their existence, that what applies to the one is almost sure not to apply to the other. Domestic animals are abnormal, irregular, artificial; they are subject to varieties which never occur and never can occur in a state of nature; their very existence depends altogether on human care; so far are many of them removed from that just proportion of faculties, that true balance of organization, by means of which alone an animal left to its own resources can preserve its existence and continue its race.

The hypothesis of Lamarck – that progressive changes in species have been produced by the attempts of animals to increase the development of their own organs, and thus modify their structure and habits – has been repeatedly and easily refuted by all writers on the subject of varieties and species, and it seems to have been considered that when this was done the whole question has been finally settled; but the view here developed renders such an hypothesis quite unnecessary, by showing that similar results must be produced by the action of principles constantly at work in nature. The powerful retractile talons of the falcon- and the cat-tribes have not been produced or increased by the volition of those animals; but among different varieties which occurred in the earlier and less highly organized forms of these groups, those always survived longest which had the greatest facilities for seizing their prey. Neither did the giraffe acquire its long neck by desiring to reach the foliage of the more lofty shrubs, and constantly stretching its neck for the purpose, but because any varieties which occurred among its antitypes with a longer neck than usual at once secured a fresh range of pasture over the same ground as their shorter-necked companions, and on the first scarcity of food were thereby enabled to outlive them. Even the peculiar colours of many animals, especially insects, so closely resembling the soil or the leaves or the trunks on which they habitually reside, are explained on the same principle; for though in the course of ages varieties of many tints may have occurred, yet those races having colours best adapted to concealment from their enemies would inevitably survive the longest. We have also here an acting cause to account for that balance so often observed in nature, – a deficiency in one set of organs always being compensated by an increased development of some others – powerful wings accompanying weak feet, or great velocity making up for the absence of defensive weapons; for it has been shown that all varieties in which an unbalanced deficiency occurred could not long continue their existence. The action of this principle is exactly like that of the centrifugal governor of the steam engine, which checks and corrects any irregularities almost before they become evident; and in like manner no unbalanced deficiency in the animal kingdom can ever reach any conspicuous magnitude, because it would make itself felt at the very first step, by rendering existence difficult and extinction almost sure soon to follow. An origin such as is here advocated will also agree with the peculiar character of the modifications of form and structure which obtain in organized beings – the many lines of divergence from a central type, the increasing efficiency and power of a particular organ through a succession of allied species, and the remarkable persistence of unimportant parts such as colour, texture of plumage and hair, form of horns or crests, through a series of species differing considerably in more essential characters. It also furnishes us with a reason for that 'more specialized structure' which Professor Owen states to be a characteristic of recent compared with extinct forms, and which would evidently be the result of the progressive modification of any organ applied to a special purpose in the animal economy.

We believe we have now shown that there is a tendency in nature to the continued progression of certain classes of varieties further and further from the original type – a progression to which there appears no reason to assign any definite limits – and that the same principle which produces this result is a state of nature will also explain why domestic varieties have a tendency to revert to the original type. This progression, by minute steps, in various directions, but always checked and balanced by the necessary conditions, subject to which alone existence can be preserved, may, it is believed, be followed out so as to agree with all the phenomena presented by organized beings, their extinction and succession in past ages, and all the extraordinary modifications of form, instinct, and habits which they exhibit.

Ternate, February, 1858

Notes

Introduction

1 Gruber, 1981, p275

Chapter 1

1 Correspondence of Charles Darwin (hereafter referred to as CCD), 6: 297
2 CCD 6: 99
3 CCD 6: 106
4 Wilson, 1970, p6
5 Wilson, pp65–67
6 McKinney, 1972, p117
7 De Beer, 1959, May 14 1856, p14
8 Browne, 1995, p513
9 Ospovat, 1995, pp73–86
10 Ospovat, 1995, pp185–186
11 *The Times*, 12 January 1857
12 CCD 6: 387
13 Berry, 2002; Wallace,1856c, pp36–49
14 CCD 6: 387 (see British Library: Addison Collection Ref: ADD 46434 for original letter)

Chapter 2

1 Wilson, 1972, p237
2 Wilson, p251
3 Wilson, p258
4 Wilson, p258
5 Wilson, p259
6 Wilson, pp69–73
7 Eiseley, p39
8 Eiseley, pp41–42
9 Eiseley, p42
10 Eiseley, p42
11 Ospovat, 1995, p2

Chapter 3

1 Darwin, 1839
2 Darwin, 1845
3 Bowlby, 1990, p60
4 Barlow, 1958, p44
5 Darwin, 1887, Vol. 1, pp10–11
6 Darwin, p86
7 Bowlby, 1990, p71
8 Barlow, 1958, p11
9 Darwin, 1887, Vol. 1, p11
10 DAR 140/3 (unpublished material in Darwin Archive)
11 Darwin, 1887, Vol. 1, p41
12 Darwin, 1887, Vol. 3, pp219–220
13 Jesperson, 1948–1949
14 CCD 1: 488
15 CCD 1: 516
16 CCD 2: 7
17 Lyell, 1881, Vol. 1, pp457–459, pp460–461
18 CCD, 1: 516
19 CCD, 2: 29
20 CCD, 2: 88
21 Lyell, 1881, Vol. 2, pp39–41
22 CCD, 2: 104
23 Darwin, 1887, Vol. 1, pp82–83

Chapter 4

1 Darlington, 1959
2 Darlington, p1
3 Darlington, p4
4 Darlington, p4
5 Darlington, p4
6 Darlington, p6
7 Darlington, p27
8 Darlington, p27
9 Darlington, p57
10 Darwin, 1887
11 Darwin, 1909
12 Darwin, 1887, Vol 1, pp83–84
13 Barlow, 1958, p120
14 Eiseley, 1959
15 Eiseley, 1979

Chapter 5

1 Darwin, 1909, pxvi
2 Blyth, 1835, 1836, 1837
3 Barlow, 1946, p263
4 Eiseley, 1979, p47
5 Eiseley, p47

6 Eiseley, p47
7 Darwin, 1909
8 Eiseley, 1979, p49
9 Eiseley, p47, pp50–53
10 Eiseley, p51
11 Eiseley, p52
12 Eiseley, p53
13 Eiseley, pp62–63
14 Eiseley, pp54–59
15 Eiseley, p64
16 Eiseley, p67
17 Eiseley, p55
18 Eiseley, p55
19 Eiseley, p59
20 Eiseley, pp59–62
21 Eiseley, p72. Original material in Matthew, 1831, p207, p308
22 Eiseley, p73, p78
23 Eiseley, p67
24 Henderson, 1958, p54.
25 Eiseley, 1979, p68
26 Desmond and Moore, 1991, pp317–318

Chapter 6

1 Gruber and Barrett, 1974
2 Gruber and Barrett, p3
3 Gruber and Barrett, p20
4 Gruber, 1981, p20, p101
5 Gruber, 1981, pxi
6 Gruber, pp23–24
7 Gruber, p298
8 Gruber, p297
9 Gruber, p287
10 Gruber and Barrett, 1974, p100
11 Gruber 1981, pp275–276

Chapter 7

1 Hughes, 1989, p411
2 *Zoologist*, 1847, p1676
3 Wallace, 1905, Vol. 1, p151
4 Wallace, p 246. See also Hughes, 1989, p409
5 Wallace, p264
6 Wallace, p406
7 Wallace, p151, p237
8 Wallace, p232
9 Wallace, p254
10 Chambers, 1844; reprinted 1969, p389
11 Chambers, p222
12 Lyell, 1830-1833, Vol.1, p123. Also Ospovat, 1995, pp16-20; Eiseley, 1958, pp38–41
13 Wallace, 1905, Vol. 1, p254
14 Wallace *My Life*. Vol 1:254
15 Hughes, 1989
16 Hughes, 1991
17 Wallace, 1905, Vol. 1, p256
18 Brooks, 1984, p18
19 Bates, 1854, Preface
20 Brooks, 1984, p19
21 Wallace, 1905, Vol. 1, p266
22 Brooks, 1984, p19
23 *Zoologist*, 1852, October 19, pp3641–3643
24 *Zoologist*, pp3641–3643
25 *Zoologist*, pp3641–3643
26 Wallace, 1905, Vol. 1, p306
27 *Annals and Magazine of Natural History*, 1848a
28 *Annals and Magazine of Natural History*, 1848b
29 *Annals and Magazine of Natural History*, 1849a
30 *Annals and Magazine of Natural History*, 1849b
31 *Annals and Magazine of Natural History*, 6, 1849b
32 *Zoologist*, 1851
33 Wallace, 1905, Vol. 1, p285
34 *Proceedings of the Zoological Society*, 1852
35 Lyell, 1991 (originally published 1830–1833), Vol. 2, p66
36 Wallace, 1853a, pp294–296, pp425–427
37 *Proceedings of the Zoological Society*, 1850, pp206–207
38 Brooks, 1984, pp42–3
39 Wallace, 1853a, p246
40 Wallace, pp210–211, pp341–342
41 Spencer, 1852
42 *Transcripts of the Entomological Society*, 1852–1853
43 Lyell, 1830–1833, pp144–147
44 *Transcripts of the Entomological Society*, 1852–1853

Chapter 8

1 Wallace, 1853a, 1853c
2 CCD 5: 214
3 Ospovat, 1995, p55; Barrett, 1987, extract from B notebook, pp17–21.
4 CCD 4: 15
5 CCD 4: 21
6 CCD 4: 20
7 CCD 4: 29

8 CCD 4: 343. Also contrast Browne, 1995, Vol 1: 513/4 with Ospovat pp201–202
9 DAR 205.5. See Browne, 1980; De Beer, 1963, pp139–140; Ospovat, 1995, p176.
10 CCD 3: 290
11 CCD 3: 293
12 CCD 3: 295
13 CCD 5: 267
14 Wollaston pp xiii–xiv
15 CCD 5: 279
16 Browne, 2003, pp517–521
17 CCD 5: 325
18 *Gardeners' Chronicle and Agricultural Gazette*, 1855
19 CCD 5: 338 2. CCD 5: 343
20 CCD 5: 345
21 CCD 5: 353
22 CCD 5: 379

Chapter 9

1 Wallace, 1854
2 Linnaeus, 1758
3 Wallace, 1905, Vol. 1, p329
4 Wallace, pp354–355
5 Wallace, 1855b
6 Brooks, 1984, pp84–96, pp403–412
7 Wallace, 1905, Vol. 1, pp354–355
8 Wallace, 1855a
9 Wallace, 1855c, pp195–196
10 Wallace, 1855a. See also Brooks, 1984, pp114–115.
11 Wallace, 1855c
12 Evenhuis, 2003
13 Darwin, 1887, Vol. 2, pp179–205

Chapter 10

1 McKinney, 1972, pp25–26
2 CCD 6: 89

Chapter 11

1 Wilson, 1970
2 Wilson, p65
3 Wilson, p79
4 Wilson, p54
5 CCD 6: 89
6 CCD 6: 99
7 CCD 6: 106
8 Hooker's May reply to Darwin: letter not found
9 CCD 6: 108
10 DAR 205.5:174.
11 De Beer, 1959, p14
12 Stauffer,1975, p8. See also Stauffer, 1959
13 CCD 6:135

Chapter 12

1 McKinney, 1972
2 McKinney, pp35–43
3 McKinney, p43
4 McKinney, p117
5 McKinney, p117
6 McKinney, p118
7 McKinney, p117; p119
8 Brooks, 1984, p236
9 CCD 5: 519
10 CCD 5: 519n
11 McKinney, 1972, p120

Chapter 13

1 CCD 6: 124; 6: 126
2 CCD 6: 143
3 CCD 6: 144
4 CCD 6: 153
5 CCD 6: 167
6 CCD 6: 166
7 CCD 6: 178
8 CCD 6: 189
9 CCD 6: 190
10 CCD 6: 198
11 CCD 6: 200
12 Wallace, 1856a
13 Wallace, 1856c

Chapter 14

1 Wallace, 1855b
2 *Zoologist*, 1856
3 Lyell, 1835, Vol. 3, pp18–19
4 Wallace, 1856a, p27
5 Wallace, pp26–32
6 Brooks, 1984, pp103–104
7 Wallace, 1856c
8 Wallace, 1856c

Chapter 15

1 Ospovat, 1995, p1
2 Ospovat, pp82–83
3 Ospovat, p192
4 Ospovat, p84
5 Ospovat, p199
6 Ospovat, p210
7 Ospovat, p62
8 Ospovat, p1
9 Ospovat, p205

Chapter 16

1 CCD 6: 217
2 DAR 205.5: 171
3 CCD 6: 233
4 CCD 6: 235
5 CCD 6: 237
6 CCD 4: 344. See also Ospovat, 1995, pp 200–202
7 CCD 6: 244
8 CCD 6: 257
9 CCD 6: 259
10 CCD 6: 266
11 CCD 6: 271
12 CCD 6: 281
13 Ospovat, 1995, p174
14 CCD 6: 297

Chapter 17

1 Wallace, 1856d
2 Wallace, 1856d
3 Wallace, 1856d
4 Wallace, 1856e
5 Wallace, 1856b
6 Brooks, 1984, p139
7 *Zoologist* (3 Nov, 1856): 5414-5

Chapter 18

1 Ospovat, 1995, p172, p173
2 Ospovat, p200
3 CCD 6: 271
4 Ospovat, 1995, p210
5 Ospovat, p210
6 Darwin, 1887, Vol. 1, pp83–88

Chapter 19

1 Beddall, 1968, p19
2 Brackman, 1980, p348
3 Brackman, p348
4 Brackman, p348
5 CCD 6: 387. Also British Library: Addison Collection ADD 46434, for original letter.

Chapter 20

1 *Zoologist*, 1857
2 Wallace, 1856b
3 *Javasche Courant*, No 83, 15 October 1856
4 *Singapore Free Press*, 20 October 1856, pg 3
5 *Singapore Free Press*, 30 October 1856, pg 3
6 *The Times*, No 22,545, 8 December 1856, pg 8; *Zoologist*, 1857
7 CCD 6: 387. British Library: ADD 46434. Also see Appendix 2
8 *The Times*, No 22,646, 4 April 1857, pg 10
9 *The Times*, No 22,671, 4 May 1857, pg 10
10 *The Times*, No 22,666, 28 April 1857, pg 6

Chapter 21

1 CCD 6: 387. Also British Library: Addison Collection ADD 46434 for original letter.
2 CCD 6: 314
3 CCD 6: 346
4 DAR 10.2: MS: 27 and 28. See also Stauffer, 1975, p250
5 CCD 6: 370
6 CCD 6: 376
7 CCD 6: 387
8 McKinney, 1972, pp118–119
9 CCD 6: 387
10 McKinney, 1972, pp124–129
11 Ospovat, 1995, p201
12 Ospovat, p205
13 Browne, 1980, p73. See also Ospovat, 1995, pp185–186
14 CCD 6: 401
15 CCD 6: 412
16 CCD 6: 422
17 CCD 6: 431
18 CCD 6: 438
19 CCD 6: 443
20 CCD 6: 437

Chapter 22

1 Brooks, 1984, pp229–231
2 Linnean Society, various dates
3 Brooks, 1984, p147
4 Wallace, 1858c
5 Wallace, 1858a
6 Wallace, 1857b
7 Wallace, 1857c
8 *Zoologist*, 1857
9 Wallace, 1857b
10 Brooks, 1984, p149
11 Brooks, p155
12 Wallace, 1857b
13 Wallace, 1857b
14 Wallace, 1857b
15 Wallace, 1857b
16 Berry, 2002, p35.
17 CCD 6: 387. See British Library: Addison Collection ADD 46434 for original letter with markings
18 Wallace, 1905, Vol. 1, pp358–359

19 Beddall, 1968, p292
20 Browne, 1980

Chapter 23

1 CCD 6: 445
2 For a critical analysis of Darwin's letters to Gray see Brooks J.L. 1984, pp204–212
3 CCD 6: 491
4 CCD 6: 494
5 British Library: Addison Collection ADD 46434. The date of writing of Wallace's letter is referred to in Darwin's reply
6 British Library: Addison Collection ADD 46434. See also Appendix 3
7 CCD 6: 498
8 CCD 7: 69
9 CCD 7: 82
10 British Library: Addison Collection ADD 46434. See comments in Darwin's second letter to Wallace regarding land bridges
11 CCD 7: 84

Chapter 24

1 Wallace, field journal, entry 126 (February 1858)
2 Wallace, field journal, entry 127 (1 March 1858)
3 Wallace, field journal, entry 71 (13–28 March 1857)
4 Wallace, field journal, entry 77 (1 April – 8 May 1857)
5 Wallace, field journal, entry 82 (April, 1857)
6 Brooks, 1984, pp179–180
7 Wallace, 1903, pp78–79
8 Wallace, 1905, Vol. 1, pp361–363
9 Berry, 2002; Wallace, 1858d
10 Brooks, 1984, p189
11 *Javasche Courant*, No 27, 5 April 1858

Chapter 25

1 Brooks, 1984, pp229–231
2 Brooks, p231
3 Brooks, p232
4 Brooks, pp232–233
5 Brooks, pp232–233
6 Brooks, p241
7 Brooks, p232
8 Brooks, p251
9 Brooks, p255

Chapter 26

1 Timeline based on the following editions: *Javasche Courant* 12 November 1856; *Singapore Free Press* 20 November 1856; *Colombo Observer* 15 December 1856; *Lloyd's List* 2 January 1857; *Lloyd's List* 9 January 1857; *The Times* 12 January 1857. See also Appendix 2

Chapter 27

1 British Library: Addison Collection ADD 46434
2 Beddall, 1968, p303
3 *The Times*, No 22, 869, 21 December 1857, pg 4
4 *Singapore Free Press*, Vol 25, No 5, 4 February 1858, pg 3
5 *Javasche Courant*, No 16, 24 February 1858 - Scheepsberigten
6 Wallace, journal, p66, entries 128 and 129
7 Wallace, 1905, Vol. 1, p366

Chapter 28

1 Brackman, 1980, pp343–344
2 Brackman, p343
3 McKinney, 1972, p139
4 Timeline based on the following: Wallace field journal, Linnean Society Library, London; *Javasche Courant*/Prof Femme Gaastra; *Javasche Courant* 5 April 1858; *Singapore Free Press* 22 April 1858; *Ceylon Overland Observer* 10 May 1858; *Indian & Australian Mail Homewards, 1857-61*(Post 43/156), Post Office Museum, Mount Pleasant, London; *The Times* 3 June 1858; Wallace Archives, Natural History Museum, London. See also Appendix 2

Chapter 29

1 Linnean Society Library, London. Record of attendance at meetings in 1858.
2 CCD 7: 107
3 CCD 7: 117
4 CCD 7: 119
5 CCD 7: 121. (Notes 5 and 6 have the same reference in the correspondence of Charles Darwin.)
6 CCD 7: 121
7 See Browne, 2003, Vol. 2, pp35–42 for comprehensive account

8 Darwin and Wallace, 1858
9 Wallace, *My Life*, 1905, Vol. 1, p365
10 Darwin, 1859

Chapter 30

1 CCD 7: 125
2 Darwin, 1887, pp462–463; see also Wallace, 1905
3 Brooks, 1984, p215
4 Slotten, 2004, pp356–364
5 Marchant, 1916, p18
6 Marchant, p21
7 Marchant, p39
8 Darwin, 1887, pp179–204. See Wallace, 1905, for Wallace's reaction to this praise
9 Northrop, 1913, p622

Bibliography and sources

ANNALS AND MAGAZINE OF NATURAL HISTORY, 1848a, July

ANNALS AND MAGAZINE OF NATURAL HISTORY, 1848b, 2d Series, 3, pp74–75

ANNALS AND MAGAZINE OF NATURAL HISTORY, 1849a, 2d Series, 5, pp156–157

ANNALS AND MAGAZINE OF NATURAL HISTORY, 1849b, 2d Series, 6, letter dated 15 November, pp494–496

BARLOW, Nora (ed.), 1933, *Charles Darwin's Diary of the Voyage of H.M.S. Beagle*, Cambridge, Cambridge University Press (reprinted, 1969, New York, Kraus Reprint Co.)

BARLOW, Nora, 1946, *Charles Darwin and the Voyage of the Beagle*, New York, Philosophical Library

BARLOW, Nora, 1958, *The Autobiography of Charles Darwin, 1809–1882, with original omissions restored*, London, Collins

BARRETT P. H. (ed.), 1987, *Charles Darwin's Notebooks 1836–44*, London, British Museum and Cambridge University Press

BATES, H. W., 1863, *The Naturalist on the River Amazons*, London, Murray

BEDDALL, Barbara G., 1968, 'Wallace, Darwin, and the theory of natural selection: a study in the development of ideas and attitudes', *Journal of the History of Biology*, 1(2), pp261–324

BEDDALL, Barbara G. (ed.), 1969, *Wallace and Bates in the Tropics; An Introduction to the Theory of Natural Selection*, London, Macmillan

BEDDALL, Barbara G., 1972, 'Wallace, Darwin, and Edward Blyth: further notes on the development of evolution theory', *Journal of the History of Biology*, 5(1), pp153–158

BEDDALL, Barbara G., 1988, 'Darwin and divergence: the Wallace connection', *Journal of the History of Biology*, 21(1), pp1–68

BERRY, Andrew (ed.), 2002, *Infinite Tropics: An Alfred Russel Wallace Anthology*, London and New York, Verso

BLYTH, E., 1835, 'An attempt to classify the "varieties" of animals', *The Magazine of Natural History*, 8(1), Parts 1–2

BLYTH, E., 1836, 'Seasonal and other external changes in birds', *The Magazine of Natural History*, 9(1), Parts 1–3

BLYTH, E., 1837, 'The psychological distinctions between man and all other animals', *The Magazine of Natural History*, 10, Parts 1–4

BOISDUVAL, J. A., 1832, *Voyage de Decouvertes de l'Astrolabe: Faune entomologique de l'Ocean Pacifique*, Paris, J. Tastu

BOISDUVAL, J. A., 1836, *Diurnes: Papilionides, pierides: Especes Generales des Lepidopteres. Histoire naturelle des insects*, Vol.1, Paris, Libraire Encyclopedique de Paris.

BOWLBY, John, 1990, *Charles Darwin: A Biography*, London, Hutchinson

BRACKMAN, A. C., 1980, *A Delicate Arrangement: The Strange Case of Charles Darwin and Alfred Russel Wallace*, New York, Times Books

BROOKS, J. L., 1969, 'Re-assessment of Alfred Russel Wallace's contribution to the theory of organic evolution', *Yearbook 1968*, Philadelphia, PA, American Philosophical Society

BROOKS, J. L., 1972, 'Extinction and the origin of organic diversity', in E. S. Deevey (ed.), *Growth by Intussusception*, Connecticut Academy of Arts and Science, pp19–56

BROOKS, J. L., 1984, *Just Before the Origin: Alfred Russel Wallace's Theory of Evolution*, New York, Columbia University Press

BROOKS, J. L., 1985, 'Development of Wallace's perceptions of biogeography, 1848–1859', *Earth Sciences History*, 4(2), pp113–117

BROWNE, Janet, 1980, 'Darwin's botanical arithmetic and the "Principle of Divergence", 1854–1858', *Journal of the History of Biology*, 13(1), pp53–89

BROWNE, Janet, 1995, *Charles Darwin: Voyaging*, London, Pimlico

BROWNE, Janet, 2003, *Charles Darwin: The Power of Place*, London, Pimlico

BURKHARDT, F. H. and SMITH, S. (eds), 1983–1994, *The Correspondence of Charles Darwin*, Vols 1–7, Cambridge, Cambridge University Press

CHAMBERS, Robert, 1844, *Vestiges of the Natural History of Creation*, London, reprinted with an introduction by J. A. Secord, 1994, Chicago, University of Chicago Press

DARLINGTON, C. D, 1959, *Darwin's Place in History*, Oxford, Blackwell

DARWIN, Charles, 1839, *Journal of Researches into the Geology and Natural History of the Various Countries Visited by H.M.S. Beagle*, London, Henry Colburn

DARWIN, Charles, 1845, *Journal of Researches into the Natural History and Geology of the Countries Visited During the Voyage of H.M.S. Beagle Round the World*, London, Murray

DARWIN, Charles, 1851–1854, *A Monograph on the Sub-class Cirripedia*, London, Ray Society

DARWIN, Charles, 1859, *On the Origin of Species by Means of Natural Selection, or the Preservation of Favoured Races in the Struggle for Life*, London, Murray, reprinted with an introduction by J. W.Burrow, 1968, Harmondsworth, Penguin

DARWIN, Charles and WALLACE, A. R., 1858, 'On the tendency of species to form varieties; and on the perpetuation of varieties and species by natural means of selection', *Journal of the Proceedings of the Linnean Society*, 3, pp46–50

DARWIN, Erasmus, 1818, *Zoonomia; or the Laws of Organic Life*, fourth American edition, Philadelphia, Edward Earle

DARWIN, Francis (ed.), 1887, *The Life and Letters of Charles Darwin, Including an Autobiographical Chapter*, London, Murray

DARWIN, Francis, 1909, *The Foundations of the Origin of Species: Two Essays Written in 1842 and 1844 by Charles Darwin*, Cambridge, Cambridge University Press

DARWIN, Francis and SEWARD, A. C. (eds), 1903, *More Letters of Charles Darwin*, London, Murray

DE BEER, Gavin (ed.), 1959, 'Darwin's journal', *Bulletin of the British Museum (Natural History) Historical Series*, 2(1), pp1–21

DE BEER, Gavin, 1963, *Charles Darwin: Evolution by Natural Selection*, Edinburgh, Nelson

DESMOND, Adrian and MOORE, James R., 1991, *Darwin*, London, Michael Joseph

DUPREE, A. H., 1968, *Asa Gray, 1810–1888*, New York, Athenaeum

EISELEY, Loren, 1958, *Darwin's Century: Evolution and the Men who Discovered It*, New York, Doubleday

EISELEY, Loren, 1959, 'Charles Darwin, Edward Blyth and the theory of 'Natural Selection', *Proceedings of American Philosophical Society*, 103(1), pp94–114

EISELEY, Loren, 1979, *Darwin and the Mysterious Mr X: New Light on the Evolutionists.* London, Toronto and Melbourne, J. M. Dent.

EVENHUIS, N L 2003, 'Publication and dating of the journals forming the *Annals and Magazine of Natural History* and the *Journal of Natural History*', *Zootaxa*, 385 (December), pp1–68

FORBES, Edward, 1846, 'On the connexion between the distribution of the existing flora and fauna of the British Isles and the geological changes which have affected their area', *Memoirs of the Geological Survey of Great Britain and Ireland*, London, Geological Society of London, pp336–432

Gardeners' Chronicle and Agricultural Gazette, 1855, 26 May, pp356–357

GEORGE, W., 1964, *Biologist Philosopher: A Study of the Life and Writings of Alfred Russel Wallace*, London, Abelard-Schuman

GEORGE, W., 1981, 'Wallace and his Line' in T. C. Whitmore (ed.), *Wallace's Line and Plate Tectonics*, Oxford, Oxford University Press

GRUBER, H. E., 1981, *Darwin on Man: A Psychological Study of Scientific Creativity* (second edition), Chicago, University of Chicago Press

GRUBER, H. E. and BARRETT, P. H., 1974, *Darwin on Man: A Psychological Study of Scientific Creativity*, London, Wildwood House

HENDERSON, Gerald M., 1958, 'Alfred Russel Wallace: his role and influence in nineteenth century evolutionary thought', unpublished Ph.D. dissertation, University of Pennsylvania, PA

HUGHES, R. E., 1989, 'Alfred Russel Wallace: some notes on the Welsh connection', *British Journal for the History of Science*, 22, pp401–418

HUGHES, R. E., 1991, 'Alfred Russel Wallace: the making of a scientific nonconformist', *Proceedings of the Royal Institution of Great Britain*, 63, pp175–183

JESPERSON, P. H., 1948–1949, *Charles Darwin and Dr Grant*, Lychnos

LINNAEUS, Carl, 1758, *Systema Naturae*, tenth edition, Stockholm, L. Salvii

LITCHFIELD, Henrietta (ed.), 1915, *Emma Darwin: A Century of Family Letters, 1792–1896*, New York, Appleton

LYELL, Charles, 1830–1833, *Principles of Geology: Being an Attempt to Explain the Former Changes of the Earth's Surface, by Reference to Causes Now in Operation*, fourth edition, London, Murray. Facsimile reprint with an introduction by M. J. S. Rudwick, 1991, Chicago, University of Chicago Press

LYELL, K. M., 1881, *Life, Letters and Journals of Sir Charles Lyell, Bart*, 2 volumes, London, Murray

MCKINNEY, H. L., 1966, 'Alfred Russel Wallace and the discovery of natural selection', *Journal of the History of Medicine and Allied Sciences*, 21, pp333–357

MCKINNEY, H. L., 1967, 'Alfred Russel Wallace and the discovery of natural selection', Ph.D. dissertation, Cornell University, NY

MCKINNEY, H. L., 1972, *Wallace and Natural Selection*, New Haven and London, Yale University Press

MALTHUS, Thomas Robert, 1798, *An Essay on the Principle of Population, as it Affects the Future Improvement of Society*, edited with an introduction by Antony Flew, 1970, Harmondsworth, Penguin

MARCHANT, J., 1916, *Alfred Russel Wallace: Letters and Reminiscences*, London, Cassell. Reprint, 1975, New York, Arno Press

MATTHEW, P., 1831, *On Naval Timber and Arboriculture*, London, Longman

NEWMAN, E., 1854, 'President's Address for 1853', *Transactions of the Entomological Society of London*, 2, pp142–154

NORTHROP, W. B., 1913, 'Alfred Russel Wallace', *The Outlook*, 105, p622

OSPOVAT, Dov, 1995, *The Development of Darwin's Theory: Natural History, Natural Theology, and Natural Selection 1838-1859* Cambridge, Cambridge University Press

PANTIN, C. F. A., 1959, 'Alfred Russel Wallace, F.R.S., and his essays of 1858 and 1855', *Notes and Records of the Royal Society of London*, 14(1), pp67–84

PANTIN, C. F. A., 1959 , 'Alfred Russel Wallace: his pre-Darwinian essay of 1855', *Proceedings of the Linnean Society of London 1958–59*, 171(2), pp139–153

POST OFFICE ARCHIVES, *HMG: India and Australian Mail Homewards 1857–1861*, Post 45/156 Daily Packet Lists of 1858, GPO Museum, London

PROCEEDINGS OF THE ZOOLOGICAL SOCIETY, 1850, Part 18, London, Zoological Society, pp206–207

PROCEEDINGS OF THE ZOOLOGICAL SOCIETY, 1852, Part 20, London, Zoological Society, pp107–110

SCHWEBER, S. S., 1980, 'Early Victorian science: science in culture', *Journal of the History of Biology*, 13(1), pp121–140

SHERMER, Michael, 1995, 'A gentlemanly arrangement: Alfred Russel Wallace, Charles Darwin, and the resolution of a scientific priority dispute', *Skeptic*, 3(2), pp80–89

SHERMER, Michael, 2001, *The Borderlands of Science: Where Sense Meets Nonsense*, Oxford, Oxford University Press

SLOTTEN, R., 2004, *The Heretic in Darwin's Court*, New York, Columbia University Press

SMITH, S., 1960, 'The origin of The Origin', *Advancement of Science*, 64(16), pp391–401

SPENCER, H., 1852, 'The developmental hypothesis', *The Leader*, March 20, London

STAUFFER. R. C. (ed.), 1975, *Charles Darwin's Natural Selection, Being the Second Part of his Big Species Book Written From 1856 to 1858*, Cambridge, Cambridge University Press

STAUFFER. R. C., 1959, 'On the Origin of Species: an unpublished version', *Science*, 130(3387), pp1449–1452

SULLOWAY, Frank, 1982, 'Darwin and his finches: the evolution of a legend', *Journal of the History of Biology*, 15, pp1–53

TRANSCRIPTS OF THE ENTOMOLOGICAL SOCIETY, 1852–1853, London, Entomological Society, pp142–154, pp253–264

WALLACE, Alfred Russel, notebooks and journals, London, Linnean Society Library

WALLACE, Alfred Russel, 1850, 'On the umbrella bird (*Cephalopterus ornatus*)', *Proceedings of the Zoological Society of London*, part 18, pp206–207

WALLACE, Alfred Russel, 1853a, *A Narrative of Travels on the Amazon and Rio Negro*, London. Reprint, 1969, New York, Haskell House

WALLACE, Alfred Russel, 1853b, 'On the Rio Negro', *Journal of the Royal Geographical Society*, 23, pp212–217

WALLACE, Alfred Russel, 1853c, *Palm Trees of the Amazon and their Uses*. Reprint, 1971, Lawrence, KA, Coronado Press

WALLACE, Alfred Russel, 1853d, 'On the habits of the butterflies of the Amazon Valley', *Transactions of the Entomological Society of London*, 7, pp253–264

WALLACE, Alfred Russel, 1854, 'The Entomology of Malacca', *Zoologist*, 13, p4636

WALLACE, Alfred Russel, 1855a, 'On the ornithology of Malacca', *Annals and Magazine of Natural History*, 15, pp95–99

WALLACE, Alfred Russel, 1855b, 'Description of a new kind of *Ornithoptera*', *Transactions of the Entomological Society of London*, 3, pp104–105

WALLACE, Alfred Russel, 1855c, 'On the law which has regulated the introduction of new species', *Annals and Magazine of Natural History*, 16(93), pp184–196

WALLACE, Alfred Russel, 1855d, 'Species notebook', London, Linnean Society Library

WALLACE, Alfred Russel, 1856a, 'On the habits of the Orang-utan in Borneo', *Annals and Magazine of Natural History*, 17(103), pp26–32

WALLACE, Alfred Russel, 1856b, letter to Stevens read out at meeting of Zoological Society, November 3

WALLACE, Alfred Russel, 1856c, 'Attempts at a natural arrangement of birds', *Annals and Magazine of Natural History*, 2d series, 18, pp193–216

WALLACE, Alfred Russel, 1856d, 'Splendours of the eastern archipelago – observations on the geology of Borneo', *Zoologist*, pp5116–5117

WALLACE, Alfred Russel, 1856e, 'Journal', entry 8a, May (unpublished)

WALLACE, Alfred Russel, 1857a, 'On the great bird of paradise of the Malays', *Annals and Magazine of Natural History*, 2d series, 20, pp411–416

WALLACE, Alfred Russel, 1857b, 'On the natural history of the Aru islands', *Annals and Magazine of Natural History*, 20, supplement, pp473–485

WALLACE, Alfred Russel, 1857c, letter to S. Stevens from Aru, 15 May (printed in *Proceedings of the Entomological Society of London*, 8, pp9193)

WALLACE, Alfred Russel, 1858a, 'Note on the theory of permanent and geographical varieties', *Zoologist*, 16, pp5887–5588

WALLACE, Alfred Russel, 1858b, 'On the entomology of the Aru islands', *Zoologist*, 16, pp5887–5888

WALLACE, Alfred Russel, 1858c, 'On the habits and transformations of a species of *Ornithoptera*, allied to *O. Priamus*, inhabiting the Aru islands near New Guinea', *Transactions of the Entomological Society of London*, 4, pp272–273

WALLACE, Alfred Russel, 1858d, 'On the tendency of varieties to depart indefinitely from the original type: instability of varieties supposed to prove the permanent distinctness of species', *Journal of the Proceedings of the Linnean Society of London (Zoology)*, 3, pp53–62

WALLACE, Alfred Russel, 1869, *The Malay Archipelago*, Oxford, Oxford University Press

WALLACE, Alfred Russel, 1903, 'Dawn of a great discovery: my relations with Darwin in reference to the theory of natural selection', *Black & White Magazine*, January 17, pp78–79

WALLACE, Alfred Russel, 1905, *My Life: A Record of Events and Opinions*, London, Chapman and Hall*

WHITE, Michael and GRIBBIN, John, 1995, *Darwin: A Life in Science*, London, Simon and Schuster

WILSON, Leonard G. (ed.), 1970, *Sir Charles Lyell's Scientific Journals on the Species Question*, New Haven, CT, Yale University Press

WILSON, Leonard G., 1972, *Charles Lyell: The Years to 1841: The Revolution in Geology*, London and New Haven, CT, Yale University Press

WOLLASTON, Thomas Vernon, 1856, *On the Variation of Species, with Especial Reference to the Insecta*, London

ZOOLOGIST, 1847, 'Culture of Trichius fasciatus near Neath', 5, p1676

ZOOLOGIST, 1851, pp3230–3235

ZOOLOGIST, 1856, letter, pp5113–5115

ZOOLOGIST, 1857, 'Proceedings of the Entomological Society: meeting held on March 2, 1857', pp5559–5560

Index